YORKSHIRE

YORKSHIRE

A LYRICAL HISTORY OF
ENGLAND'S GREATEST COUNTY

RICHARD MORRIS

WEIDENFELD & NICOLSON

First published in Great Britain in 2018
by Weidenfeld & Nicolson
an imprint of The Orion Publishing Group Ltd
Carmelite House, 50 Victoria Embankment
London EC4Y 0DZ
An Hachette UK Company

1 3 5 7 9 10 8 6 4 2

A CIP catalogue record for this book is
available from the British Library.

ISBN (hardback) 978 0 297 60943 8
ISBN (ebook) 978 0 297 60944 5

Typeset by Input Data Services Ltd, Somerset

Printed and bound in Great Britain
by Clays Ltd, St Ives plc

MIX
Paper from
responsible sources
FSC
www.fsc.org   FSC® C104740

www.orionbooks.co.uk

To Jane
Whose Yorkshire this is

CONTENTS

ILLUSTRATIONS

FIGURES

PLATE SECTION

While every effort has been made to fulfil requirements with regard to reproducing copyright material, the author and publisher will be glad to rectify omissions at the earliest opportunity.

INTRODUCTION
AVALONIA TO BEMPTON CLIFF

M y mother was dead for twenty years when a new photograph of her turned up. It was cut from a postcard, small, scratched, a bit foggy on one side (Fig. 1). It had been hidden in plain sight, tucked away behind another picture in an album. There she was, about two years old, beside two older children on the front seat of an Edwardian car. Behind her were two ladies. One was youngish, clad in a summer dress with a scarf tied over her hair; next to her sat an older woman cradling an infant and wearing a high-crowned, broad-brimmed hat that looked a bit like the cloud from a nuclear explosion.

We scanned the picture, enlarged it and looked again. Elsie, my mother, sits behind the wheel, staring intently at the photographer, lips

1. Elsie Wearne (front left), her mother, Emily (rear right) and grandmother Hannah Smith in a beach photographer's car, Scarborough, August Bank Holiday 1911

slightly parted. A floppy ribbon is tied in her hair. The lady under the hat is Hannah, my great-grandmother. She was born near the West Riding town of Skipton in 1858 and had begun work as a servant by the age of eleven. The infant in her arms is Edna Mary, Elsie's younger sister, no more than six months old. Next to Hannah is her daughter, Emily Wearne, my grandmother. We do not recognise the other children.

The car seems to have been used as a photographer's prop.[1] Certainly, neither Hannah nor Emily was in a position to own a car. Hannah's father began as a labourer – the lowest of the low in the Victorian working class – and Emily was married to a blast furnace keeper. This said, their clothes suggest social aspiration, and Emily's head scarf hints at Edwardian fashion for lady motorists.

Where was the photograph taken? There is an open-top tram in the background. Its helical stairway can be made out on the left and the pantograph for collecting power from the overhead line is visible just behind the lamp post. The arrangement of windows shows the tram to be one of those built by the Brush Electrical Engineering Company of Loughborough. Sand in the foreground indicates a beach, the raised road and railings suggest a promenade and the tram and nearby buildings point to somewhere urban. So, this is a seaside resort.

The words painted in the long rectangular panel just above the tramcar's wheels tell us where it was. The photograph is not sharp enough to let us read them all, but the last word can be made out as 'Company' and the first is quite long, maybe ten or eleven letters. Comparison with transport archive photographs confirms the wording as 'Scarborough Tramways Company'. This fits: Scarborough used Brush trams from 1904, and my mother's family lived in a mining village about thirty miles up the coast.

The print's genre gave clues to the occasion. The spread of railways in the 1840s and introduction of Bank Holidays in 1871 fostered the working-class family day trip, for which the photograph became a memento. From the later 1880s simplified and portable processing enabled professional photographers to work on beaches and promenades.[2] Beach photographers were alert to the commercial opportunities of group photographs and works outings. Their photos were often produced in the postcard format, which is what we have here. The overlapped ages of Elsie and Edna, the signs of cool weather,[3] and the likelihood that this was a Bank Holiday together point to a date: Monday, 1 August 1910.

Having solved the picture, the picture solved a problem. Yorkshire's selfhood is more like that of an empire than a county. If you set out to write about an empire, where do you begin? The question had been niggling. Possibilities were legion. Yorkshireness, say: if it exists, what is it and how was it formed? Or dialect ('T''laadest sharters often hasn't mich on their stalls'[4])? Or what about Yorkshire and popular culture: Sooty and Sweep, *The Good Companions*, post-punk revival bands – a list soon grows. Aback of my list was a sense that a clear starting point was out there somewhere, but no idea what it was other than a feeling that I would know when I met it. And now I had. Scarborough.

The 'borough' in Scarborough comes from 'burg', which means stronghold. According to the Icelandic *Kormáks Saga* Scarborough's burg took its name from the tenth-century Viking leader Þorgils Skarði.[5] 'Skarði' is an Old Norse word for 'hare-lipped'. This has been understood as a nickname, but the site of Thorgils's eyrie on the promontory between two bays, where the later castle stands, invites another explanation: the outline of the coast at this point resembles a cleft lip. Nor was Thorgils the first to use the site. Six hundred years earlier the Roman army put an observation tower there to watch the coast, and there was a prehistoric presence before that.[6]

Later, Scarborough evolved from market town to fashionable spa, bathing place for London's well-to-do, fishing port and holiday resort. Each stage left something of itself. Regency elegance, high-Victorian finery and twenty-first-century leisure kitsch mingle round a medieval street plan. In South Bay the town is terraced up the hillside like an amphitheatre, the overlooking grey-stone castle likened to a baby Gibraltar.[7] Hollows and valleys give room for municipal gardens and opportunity for stylish ironwork bridges. Scarborough is steeped in memories. In autumn retired couples sit on benches looking out to sea, and the long, these days mostly empty, sinuous platforms at the railway station evoke crowded excursions from the West Riding. In 2015 Scarborough was still one of the five most visited places in England, although alongside the glam and nostalgia were signs of fragility – empty premises, discount stores and a lot of cash-for-gold shops.

We shall return to Scarborough, and to the neoclassical station whence trains still run to Hull through the peace of the Yorkshire Wolds. Winifred Holtby drew the line on the map she sketched of her imagined South Riding.[8] Today much of it is single track, running along Yorkshire's Jurassic coast to Filey, past mighty cliffs to Bridlington, winding through downland to Driffield and across the hummocky plain of Holderness. And here is the nub: no other English shire contains so many different landforms (Plate 1). Yorkshire is mountains, caverns, plains, precipices, chalk downs, valleys and vales, estuaries, marshland, peat bogs and upland heath. They result from happenings long ago and far away. And Scarborough is where the world was introduced to what the happenings were (Plate 2).

Understanding of rocks rests on pioneering work by William Smith (1769–1839), an Oxfordshire-born civil engineer who worked as a canal and mineral surveyor and based himself in Yorkshire from 1824. Opportunities to survey canal routes and observe excavations and mining enabled Smith to work out the order in which layers of rock had been laid down. In doing so he realised that rocks containing identical fossils must be of the same age, while rocks containing different fossils must have been formed at different times even if they looked alike. This breakthrough enabled him to identify strata which had previously been confused, and to put them into sequence. And that in turn allowed him to correlate strata and map them across the country.[9]

Smith's map enabled the landed aristocracy to anticipate what natural resources lay beneath their estates. It thus had repercussions for the distribution of wealth and structure of British society. In 1828 Smith entered the service of one of them, Sir John Johnstone (1799–1869), lord of the estate of Hackness, congenially situated a little inland from Scarborough. Johnstone was young, active, progressive-minded, and president of the Scarborough Philosophical Society. In 1828 Smith became Johnstone's land steward. Their association led to the building of a museum of geology in Scarborough. It was the first purpose-built geological museum in the world. Johnstone contributed stone for the structure, which was designed as a classical rotunda and opened in 1829. The fossils and rocks were exhibited in tiers of inward-facing curving cabinets, viewed from galleries and organised in ascending stratigraphic succession. Members of the public thus met the specimens in the same order as they occur in Yorkshire's cliffs: oldest below, youngest above.

Smith's obituarist in the *Scarborough Herald* described him as warm-hearted, cheerful and instructive. 'No man,' he wrote, 'however great his talents, ever spent an hour in his society without being enriched in knowledge, and inspired with a new ardour and energy in the pursuit of truth.'[10] Smith would have been captivated by the theory of plate tectonics, which lies aback of Yorkshire's geodiversity. The earth's shell (lithosphere) is on the move. The lithosphere consists of seven or eight sections (tectonic plates), each around seventy-eight miles thick, afloat on a ductile mantle. Earth history, at its simplest, is the story of their movement, fragmentation, reshaping and recombination. When

Yorkshire's oldest rocks were formed nearly 500 million years ago they were over 3,000 miles south of the equator, on the edge of a micro-continent known as Avalonia. Since then, Yorkshire-to-be has been travelling at the pace of a growing fingernail, inching around the globe on a voyage that – so far – has covered 6,800 miles. Along the way it has had adventures, been squeezed, tugged this way and that, and passed through a succession of climatic zones and environments. At different times Yorkshire has been desert, delta, swamp and ocean floor. At each stage new rocks have been added, and since the lithosphere interacts chemically and physically with the atmosphere and with water the rocks have undergone change.

Yorkshire's oldest rocks were later changed to slate and folded under pressure caused by continental collision. They are largely concealed, but peep out, often upended among later formations, near Ingleton and Malham. As landmasses moved, Yorkshire-to-be was uplifted, then worn down, and from 410 million years ago became part of a desert. Fifty million years later the area was close to the equator and most of the future Dales lay on the bed of a tropical sea fringed by reefs. The water teemed with mats of algae, filter-feeding brachiopods, molluscs with shells, corals, aquatic invertebrates, stem-like creatures with waving tentacles. Skeletal fragments of marine organisms formed a chalky mud that in due course changed to the limestone of the Yorkshire Dales. Joints in the limestone caused by shrinkage and later earth movements provided starting points for caves and potholes.[11]

Around 320 million years ago the warm, shallow limestone-forming sea began to interact with a river delta covering half of Yorkshire. The delta was fed by fast-flowing rivers running off granitic uplands which stood to the north. As the uplands eroded their debris was milled to sand and grit by wind and water and carried away by rivers. At the delta-front the sands sank, amalgamating with loads that cascaded from other channels to form great horizontal sheets. At other times, forests and swamp grew over the delta, leaving plant remains which later became coal. As rivers transferred rocky debris from the interior to the sea, the earth's crust rose where mountains were removed while the seabed sank under the transferred load. The sea level thus rose and fell and over tens of millions of years the processes alternated, their cyclical stages witnessed today in the step-like hard/soft bands of limestone, shale and sandstone that run along the sides of northern dales.

Gritstone country between Sheffield and Swaledale covers much of the Pennines. Adjoining Coal Measures make a gentler landscape in the Pennine foothills. These rocks are the foundation for popular York-shire – classical town halls and Wesleyan chapels, habitat of Brontës and Compo, viaducts, mines, moors, *Remains of Elmet*. How strange, then, that the unity they supply comes from opposites. Gritstone is formed of particles of rock and minerals: no fossils, just numberless grains of quartz, glassy remnants of pervasive sterility. In coal, on the other hand, is a vivid record of life. The atmosphere at this time was richer in oxygen than today, enabling plants, tracheal-breathing arthropods and insects to attain sizes that now seem prodigious. Club mosses grew taller than mill chimneys, centipedes longer than a man rippled across swamps and dragonflies the size of falcons skimmed between trees in woodlands that extended for hundreds of miles.

All the world's land masses fused around 280 million years ago. The result was a supercontinent, Pangea. The collision produced earth move-ments, faults and mountain building. Like wrinkles on a carpet caused by dragging heavy furniture, the collision had distant effects. One of them was to push the hitherto level beds of limestone, gritstone and coal into an arch-like hump – the Pennine Anticline. This upfold was much taller than now. Over the next 50 million years it was worn vir-tually to a flatland, divorcing the coalfields of Yorkshire and Lancashire and exposing the gritstone. Ninety-five million years ago the plain was back under water as sea levels rose to their highest in Yorkshire's history. The hills we see today – the Yorkshire Dales – are the result of vibra-tions from another continental collision, this time between Africa and Europe, just 16 million years ago, when the surviving Pennine fabric was again hoisted to form an area of fairly level high ground, since dis-sected by rivers and glaciers leaving remnant hilltops, like Pen-y-ghent, Whernside and Ingleborough.[12] Another effect of faults and fractures was the creation of pathways for hot saline solutions that forced their way along fissures and formed veins of lead ore and other minerals as they cooled.

To return to the story: for the next 70 million years Britain was again a desert. Sand dunes collected along the eastern flank of the recently lifted Pennines. Thirty-five million years after the birth of Pangea the dunes were flooded by the Zechstein Sea, a shallow water body with a western coastline along the Pennine foothills that extended across northern Germany to Poland 260–248 million years ago.[13] The sea was nearly landlocked and its water became hypersaline. Fluctuations in world climate and ocean level caused the sea to flood in some periods and dry up in others.[14] When the sea shrank, crystallised salts such as gypsum and halite were left encrusted in the empty basin. At Boulby, between Redcar and Whitby in East Cleveland, a several-hundred-foot thickness of Zechstein salts is mined at a depth of nearly three-quarters of a mile.[15]

Another legacy of the salty sea was the family of dolomitic limestones infused with magnesium that developed from offshore reefs and shoals. Among them are stones of the Cadeby Formation, which show delicate variations in colour from silvery grey through white to a pale buttery yellow.[16] When such rocks are first exposed they are quite 'sappy' and readily worked, but then harden in air. They are thus ideal for building. The Romans used them to build much of their fortress at York, and named a nearby town after the limestone's properties.[17] Medieval masons used stone of the Cadeby Formation for fair-faced walling, tracery and sculpture. York Minster, Selby Abbey, parts of Beverley Minster, Conisbrough Castle and hundreds of local churches are among the structures built of it. In the fifteenth century it was shipped away to royal projects like King's College Chapel in Cambridge and St George's, Windsor; later still it was used for the Houses of Parliament. More locally, craftsmen working on York Minster's chapter house in the 1280s carved limestone from Europe's Dead Sea into images of life – the foliage of trees and plants that reflected the rural world around them. Among the leaves and fruits on the chapter house walls are hawthorn, hazel, oak, ivy, buttercups and maple. Hedgerow birds and squirrels peep out of the branches. At the entrance, there is a motto: *Ut rosa flos florum, sic est domus ista domorum* – 'As the rose is the flower of flowers, so is this the house of houses'.[18]

The Zechstein sea floor tilted downwards, like the side of a shallow dish. After the sea's death, desert sand and dust accumulated in its hollow, forming red sandstone and marl that underlie the Vale of York.[19] The sandstone masks the limestone save along the sea's former shore – the dish's lip, so to speak – where the Cadeby Formation and its relatives outcrop as a north–south ridge just east of the Pennine flank. We shall meet these stones again, and their landscape, and the roads that took advantage of the ridge.[20]

By 210 million years ago Yorkshire-in-the-making had moved north to a point somewhere between modern Algeria and southern Spain. Pangea began to break up. As it did so sea levels rose. Much of England was drowned and a geologically productive period lasting 65 million years ensued. It was in this period that north-east Yorkshire's rocks were laid down in subtropical seas and deltas. Among them were limestones derived from coral, lime-rich sandstones, shales and ironstones. They form the North York Moors, and the nearby Howardian Hills.[21]

Yorkshire's Jurassic landscape has a personality all of its own. This is partly because the rocks were laid down in an undersea depression that was partitioned from the rest of Yorkshire; it also has to do with the way in which the moorland plateau and neighbouring hills took shape. The rocks were laid down flat, but at a later stage – a mere 25 million years ago – the area was uplifted and slanted by earth movements. In places the uplift brought lower strata to the surface, as in the Cleveland Hills, where layers of ironstone were exposed side on to await exploitation by industrial pioneers.[22] The tilt can also be seen in the angled, flattish tops of knolls along the southern edge of the Moors. Locally known as 'nabs', they give the range its name: the Tabular Hills.

The best place to contemplate Jurassic rocks is where William Smith studied them in the 1820s: in the tall sea cliffs between Huntcliff and Scarborough.[23] The cliffs and foreshore rock platforms enable us to look at bands not visible at the surface, and with them the fossil remains of life that teemed in the warm Jurassic sea: proto-oysters and mussels, crinoids, fish, bones of marine reptiles like plesiosaurs and ichthyosaurs. In the mid-Jurassic, around 180 million years ago, tree trunks, branches, plants and pollen washed into the sea by rivers give glimpses of inland Yorkshire's vegetation – abundant ferns, conifers, horsetails, cycads – while tree rings give a record of contemporary seasonality and the Mediterranean character of the climate that then prevailed.[24] When the sea receded and the region became coastal, dinosaurs grazed and hunted along the margins of its estuaries. Their footprints are a common sight in rocks between Scarborough and Ravenscar.

We recall that William Smith used fossils to identify strata and to work out the order in which they were laid down. His methodology was made possible by two conclusions: that fossils are the remains or marks of past life, and that when creatures or plants become extinct they never come back. These observations seem unexceptionable today; in Smith's day they were ground-breaking. For centuries, travellers, naturalists and theologians had agreed that fossils resembled living things, and some, like the Dominican friar Albertus Magnus (c. 1193–1280) had correctly ascertained them as organic forms that had been replaced by sediments.[25] However, there were intellectual hindrances to grasping the historical significance of fossils. One was the biblically derived view that the planet was only a few thousand years old, and that its rocks had been created by God. Another was the Aristotelian idea that the world was inhabited by a stable quantum of species: since fossils did not exactly correspond with anything living, the concept of fixity precluded explanations involving species that had disappeared. Conjecture instead centred on fossils as products of mysterious formative powers that operated inside rocks. Edward Lloyd (1660–1709), keeper of the Ashmolean Museum, saw them as results of 'moist seed-bearing vapours' that had penetrated the earth.[26] Robert Plot (1640–96), professor of chemistry at the University of Oxford, had the formation of crystals in mind when he attributed different fossil species to the behaviour of various 'salts'. Ammonites, for instance, 'were most probably formed either by two Salts shooting different ways, which by thwarting one another make a helical Figure, just as two opposite Winds or Waters make a Turbo ... or else by some Simple, yet unknown Salt'.[27]

Fossils invited folkloric explanations. Sharks' teeth were likened to 'tongue stones', belemnites to thunderbolts, fish teeth were serpents' eyes, the bivalve Gryphaea was a devil's toenail, and segments of crinoids were beads carved by St Cuthbert. Michael Drayton toyed with other resemblances, such as fossil sea urchins and nodules which appeared to him as 'stones of spherick forme' that 'bullets might be nam'd'.[28] Probably the most characteristic fossil of the Yorkshire coast is the ammonite, a swimming spiral-structured mollusc that ranged in size from a coin to the wheel of a truck. Drayton likened them to 'serpents ... That in their natural Gyres are up together rold'. According to legend, ammonites were the remains of snakes that had been caught, beheaded and turned to stone by the seventh-century abbess Hild (614–80). The snakes represented evil and Hild purged them to cleanse the site of her monastery (Plate 7).[29] The origin of this story is unknown, but it was already old when the Tudor topographer William Camden recounted it in his *Britannia* (1586). Walter Scott popularised the tale in *Marmion*:

> When Whitby's nuns exalting told,
> Of thousand snakes, each one
> Was changed into a coil of stone,
> When Holy Hilda pray'd:
> Themselves, within their holy ground,
> Their stony folds had often found.[30]

Marmion was begun in 1806 – the very time that William Smith was getting to grips with the significance of fossils for earth history.

Can a legend be traced to a fossil? Sockburn is an ancient settlement just a few yards outside Yorkshire in a meander of the River Tees. It is one of several northern places said to have been laid waste by a *wyrm*.[31] Wyrms were big, violent, venomous wyvern-like creatures given to taking sheep and children, uprooting trees, destroying crops and causing general mayhem. Tales of them, and of the heroism of men who confronted and killed them, were recorded at least from the twelfth century. The Sockburn Worm was reportedly dispatched by a local knight, Sir John Conyers, whose family we shall meet in Chapter 4. A wyrm's details would fit well with those of a pliosaur (Fig. 2).[32]

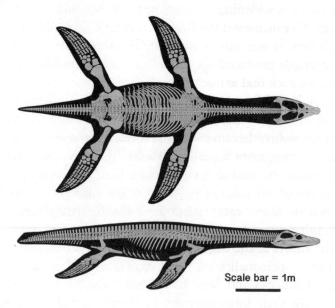

Scale bar = 1m

2. Reconstruction of the pilosaur Rhomaleosaurus

There are two, maybe three more scenes in this strange eventful history. Dinosaurs, real wyrms, flourished into the period that lasted 35 million years when myriads of microscopic shells settled on the floor of a warm sea to form the chalk of Yorkshire's Wolds. The sea was vast. Between 100 million and 65 million years ago the great land masses of Laurasia and Gondwanaland were breaking up and sea levels reached their highest since 100 million years before Yorkshire's masonry was begun. One of the factors that contributed to widespread inundation was a high level of carbon dioxide, which kept continental interiors and polar regions warmer during winters and meant that little water was sequestered as polar ice.[33] Huge tracts of chalk seabed accordingly formed. When the seabeds were uplifted they became landscapes of continental spread, stretching from eastern and southern England to the Baltic, 'through the Champagne plains on the northern edge of the Côte de l'Île-de-France in the Paris Basin to the chalk hills of Crimea, Kazakhstan and the Judean Desert'.[34] There is more chalk in Yorkshire than can be seen at the surface; chalk extends under the plain of Holderness, where glacier-borne boulder clay was later dumped across it by melting ice.

Between 65 million and 2 million years ago Britain drifted northwards from a point somewhere near modern Madrid to the latitude it occupies today. A widening gap between America and Europe became the Atlantic Ocean, moved the British land mass to the east and formed the North Sea. It was also accompanied by volcanism, and about 56 million years ago produced the Cleveland Dyke, a narrow intrusion of black volcanic rock that extends from Galloway in south-west Scotland to the North York Moors.

The end of this period saw an onset of warm–cold climatic alternation. Such swings became marked during the last 2 million years, when science recognises around seventeen of them, and at least three occasions when the cold deepened so far that ice sheets formed. The first glaciation, about 450,000 years ago, covered the entire country. The last reached its fullest extent 18,000 years ago.[35] In that instance the higher Pennine summits and North York Moors remained clear, but it is ice and the effects of the deglaciation that followed that give Yorkshire so much of its personality. More on this later, but to give examples, the rounded hillocks and hollows on the western side of the Vale of Mowbray take their intimacy from the way in which moraines and drumlins were left by confluent glaciers that had been flowing down the Vale and out of Wensleydale. It was glaciers that gave the Dales their characteristic steep-sided, flat-floored profiles, and ice over the North Sea and Holderness that dammed a vast lake in what is now the Vale of York. The Vale of Pickering, today a gently undulating plain, is the bed of another proglacial lake, which at its fullest extent had shores extending for sixty-five miles. The Nidd gorge at Knaresborough (which is a bit like part of the Dordogne accidentally left in the wrong place) was created by a torrential spillage of escaping meltwater. In 1831 the floor of an overflow channel at Newtondale in the North York Moors enabled George Stephenson to plan a railway that would connect the port of Whitby with Yorkshire's interior.[36] Quiet villages at the foot of the Wolds north of Beverley trace part of Yorkshire's coastline about a million years ago, when there was a line of chalk cliffs south of Bridlington that ran almost twenty miles inland from the present shore. During the last glacial period and for several thousand years afterwards there was no Yorkshire seaside at all, as the sea level was low and a continuous plain stretched to what is now Denmark. It is along the ancient inland coastal contour that the Scarborough–Hull railway runs.

The chalk cliffs are now on the central Yorkshire coast, jutting into the North Sea at Flamborough. Bempton is the place to see them. The train from Scarborough will take you there. As it dwindles, there is a walk from the late classical station house with its pillared porch. There are mallard on the pond; go past the church, through the village and along Cliff Lane, with its view of a derelict Cold War early warning site that looks like a handful of litter from a giant's pocket.[37] Then you are there, and there the sea cliffs are, six towering miles of them. In places the flat-bedded bands of chalk rise to 440 feet, providing ledges and crevices for nests of several hundred thousand seabirds. Kittiwakes, puffins, and the only breeding colony of gannets on the British mainland are among them. When wind blows from the east the gannets glide along the clifftop on long straight wings, the pale straw-coloured plumage of the head and neck and amethyst gaze so close. When gannets plunge-dive for fish, wings retracted, they enter the water like torpedoes. In autumn most of them migrate southwards. Gannets dislike flying overland and accordingly cross the Bay of Biscay, follow the coast of Spain, and continue southwards down Africa's Atlantic coast. They spend the next four months in the area where Yorkshire's Coal Measures were laid down 300 million years ago. In January they begin to return.

THE AINSTY AND YORK

+ Edweard cyngc gret Tostig eorl and ealle mine þegenas on
Eoferwicscire freondlice . . .¹

This is the earliest written mention of Yorkshire. It is the opening
of a royal writ, set down in the first half of the 1060s. In today's
English it reads: 'Edward the king greets earl Tostig and all my thegns
of Yorkshire in friendship . . .'²

Eoferwic (probably spoken something like 'Everwik') was the Old
English name for York. Danes or self-identified Danes ruled York and
its region from the later ninth century to the middle of the tenth. In
their speech Eoferwic slipped to *Jórvík* (say it 'Yorveek'), and the name
referred to a kingdom as well as the city.³ The kingdom became one
of England's shires when kings of Wessex took hold of it in the tenth
century. *Scír* was an Old English word denoting an area of responsibil-
ity. Tellingly, it was often hitched to other terms that had to do with
district and sway, like *scírgerefa*, 'shire reeve', whence sheriff, an official
responsible for safeguarding the king's interests.

Twenty years after Edward's writ William the Conqueror commissioned a survey to describe the resources of England's towns and
manors and learn their taxable value.⁴ A century later this was known
as the Domesday Book.⁵ Domesday's Yorkshire pages reveal a govern-
mental region in administrative working order, with three county-size
subdivisions known as *ridings* (from Old Norse þriðjungr, 'a thirding'),
and smaller administrative districts called *wapentakes*.

One of the wapentakes in the West Riding was called The Ainsty,
which lay west of the city of York and was bounded by the rivers Ouse,
Wharfe and Nidd. York claimed authority over Ainsty at least from
the thirteenth century, and between the mid-fifteenth century and the
reform of municipal corporations in 1836 the area was directly under the

city's jurisdiction. After 1836 Ainsty was returned to the West Riding. It has since disappeared as a unit of local government, but its name lives on in trading estates, pubs and streets.

1

MY WORLD BEGINS

From Norfolk, in November 1782, I passed thro' Lincolnshire into Yorkshire; my native county; — where I spent six months; —principally in observing and registering its Rural Economy: —a task I was better enabled to perform in so short a time, as my early youth was spent among it . . .

William Marshall, 1788

. . . we entered the great county of York, uncertain still which way to begin to take a view of it, for 'tis a county of very great extent.

Daniel Defoe, 1724

Carlin How is on the Cleveland coast between Saltburn-by-the-Sea and Whitby (Fig. 3). It isn't much; a bit less than a village, a touch more than a hamlet: a few terraces of workers' housing, a newsagent, a pub, chip shop, hairdresser's, and what's left of the Skinningrove iron- and steelworks across the main road. But there are family connections and the name is interesting. Eight hundred years ago everyone knew what *Kerlinghou* meant. *Kerling* is Old Norse for a hag or witch; 'how' comes from Old Scandinavian *haugr*, 'hill'. Carlin, then, has lost its 'g', and Carling How meant something like 'hill of the witches'. No one told me this when I stayed there as a child in the early 1950s, which was just as well, and only years later did I hear the name of Alfred Myers.

Myers was twenty-four when my mother was born in Carlin How in 1908 (Fig. 4). He lived nearby in a two-up, two-down brick house on Steavenson Street, and like most of his neighbours he was an ironstone miner.[1] East Cleveland is riddled with tunnels and mines – for alum,

3. *Skinningrove and Carlin How, c. 1911*

4. *The Myers brothers in the 1900s. From left to right: Henry, William,
Alfred, George and Tom*

jet, whinstone, potash, coal, but above all ironstone. Exploitation of this vast treasure in conjunction with Durham coal and bulk sea transport triggered one of the fastest regional transformations the world has seen. Accessible deposits of ironstone were recognised in Skinningrove valley below Carlin How in the 1840s.[2] The main bed was found in 1850; the first of Middlesbrough's blast furnaces was lit in 1851; by 1865 nearly fifty mines were working across Cleveland and railways had been built to serve them. In less than a lifetime Middlesbrough burgeoned from a lone marshland farm to a complex of iron-working, steel-making and shipbuilding. In 1899 a journalist found nothing in Middlesbrough, 'not even a stone', that was more than a century old. Although he was a Yorkshireman writing about Yorkshire tradition, he found this newness thrilling: Middlesbrough epitomised modernity in its purest form.[3]

Cleveland mines and Teesside industries needed more labour than local villages could provide. The new jobs accordingly attracted incomers. Some came from other mining areas. Alf Myers's father, Benjamin, had been a collier in Bradford. Tom Wearne, my grandfather, came to the ironworks with a group of tin and copper miners from Cornwall. Tom married Emily Smith, the daughter of an agricultural labourer whose family had moved up from Northamptonshire a few years before. Among their friends were the Larks, who were from Norfolk. At the start of the twentieth century this corner of Yorkshire was thronged with people who had come from somewhere else.

By 1914 Alf Myers had risen to the position of mines deputy. His duties included managing a district of the mine, keeping its roadways in order, setting props, checking ventilation and looking to the safety of the mine and its workforce. Diligent and thoughtful, he attended Carlin How's Wesleyan preaching room, taught in the Sunday school, sang in the choir and joined the Independent Labour Party.

On the evening of Friday, 15 December, a German naval force sailed from Wilhelmshaven. Its purpose was to draw out and destroy units of the British fleet by making surprise raids on east coast towns. One group of warships made for Hartlepool; a second, consisting of the battlecruisers *Derfflinger* and *Von der Tann*, and the light cruiser *Kolberg*, headed for Scarborough. The two cruisers could fling explosive shells about a foot in diameter for some miles, and soon after daybreak on the 16th they began to do so. One of those caught in the bombardment was a sixteen-year-old boarder at Queen Margaret's School called Winifred Holtby. She described what followed:

I went down to breakfast in high spirits. There was an end-of-termy feeling in the air, and breakfast was at 8 a. m. I was sitting next to Miss Crichton, and I distinctly remember she had just passed me the milk, and I was raising my first spoonful of porridge to my mouth.

I never tasted that porridge! Crash! Thu-u-d! I sat up, my spoon in the air, all the nerves in my system suddenly strung taut, for the noise was like nothing I had heard before – deafening, clear cut, not rumbly – as though a heavy piece of furniture had crashed in the room over-head. I looked at Miss Crichton, saying with a laugh, 'Hello! Who's fallen?' when the look on her face arrested me. She was deathly white and with fixed eyes was looking towards Miss Bubb . . .

I was about to speak, when Cr-r-ash – a sound more terrific than the first – and then all the windows danced in their frames; each report was doubled – first a roar and then an ear-splitting crash as the shell exploded. Then someone whispered 'guns'. The word, like magic, passed from mouth to mouth as we sat white-faced but undismayed, with the uneaten food before us. Another crash and . . . a steady voice brought us to our senses.

'Lead out to the cloak-room and wait there.' . . .

Miss Bubb appeared on the stairs. . . She was our saviour. And yet the words she said were so absurdly familiar and commonplace.

'Put on your long coats, tammies and thick boots; we are going for a walk in the country till it is over.'

We dressed and started . . . Just as we got through the gate another shell burst quite near, and 'Run!' came the order – and we ran. Ran, under the early morning sky, on the muddy, uneven road, with that deafening noise in our ears, the echo ringing even when the actual firing stopped for a moment . . .

Over the town hung a mantle of heavy smoke, yellow, unreal . . . Round the corner leading down to the Mere we ran – now all puffing. Someone was down; with a bang they fell full length on the road and lay winded; then somebody picked her up and they ran together.

In an instant's pause I looked round. I heard the roar of a gun, and the next instant there was a crash, and a thick cloud of black smoke enveloped one of the houses in Scamer Road; a tiny spurt of red flame shot out. Then I was swept down the hill . . . Where the road joins at the foot of the hill we hesitated a second; we were moving to the level crossing, when a shell struck the ground some 50 yards away, throwing up earth and mud in all directions. 'Back, back!' came the cry, and we turned and ran with dragging feet along the Mere path . . .

We crossed the line into the Seamer Valley. Along the road was a stream of refugees; there was every kind of vehicle, filled to over-flowing with women and children; yes, and men too. I saw one great brute, young and strong, mounted on a cart horse, striking it with a heavy whip, tearing at full gallop down the road, caring nothing for the women and children who scrambled piteously out of his path ... There was one particularly touching old couple, tottering along side by side ...

We paused at the foot of the hill that leads to Seamer to rest for a moment, for shells had been bursting not far from the top, and we knew that when we were half-way up we must run for our lives; all our strength was needed for that, so we stood for a moment and watched the living stream sweep past ...

Some of the girls found four tiny mites, half-dressed and almost mad with fear, yet not understanding in the least why. They had lost their mother, so we took them with us; some put coats round them and carried them. At the top of the hill we found their mother. The poor thing was almost wild with joy when she saw her 'bairns' safe and sound.

Just outside Seamer we sat down, tired out. As we sat, new comers came with dreadful tales. 'The School was shattered' – (two mistresses had stayed in!) – 'The Grand Hotel was in flames' – 'The South Cliff lay in ruins'– 'The Germans had landed ...'

In fact, the German cruisers were steaming north to bombard Whitby. The thump of their gunfire was now clearly heard in Carlin How. Back on the outskirts of Scarborough:

Some of the servants came up and told us Miss Fowler was on the road with our breakfast. Our breakfast! At this awful moment they had stopped to get chocolate, dates, and biscuits, parcels of which they had ready in case of an emergency. How good those biscuits were, eaten as we sat by the side of the road and shared them with other refugees.[4]

Winfred Holtby did not live to see publication of *South Riding* (1936), her great novel of Yorkshire and interwar England. She died aged thirty-seven; *South Riding* was seen into print by her literary executor, Vera Brittain, whose own *Testament of Youth* (1933) grappled with questions that now confronted Alf Myers.

In 1914 all members of Britain's armed forces were volunteers.

Kitchener's drive to encourage voluntary enlistment was strengthened by the raid. The eldest victim at Scarborough was sixty, the youngest, fourteen months. 'A German crime,' roared one recruiting poster. 'Remember Scarborough,' yelled another, whereon a frowning, sword-wielding Britannia gestured to a burning seaside town before a crowd of willing men. A third poster depicted a child holding a baby next to the wrecked 'home of a working man' in which four had died. 'Men of Britain! Will you stand for this?' it asked.

Scarborough, Whitby and Hartlepool enabled Britain's government and press to portray Germans as barbarians. The *Daily Mail* published an editorial entitled 'The mark of the Hun', which said that 'All paper restrictions on the conduct of warfare went into Germany's waste-paper basket the moment war was declared.'[5] Myers did not agree with the *Mail's* conclusion that laws of war were worthless, or with the official suggestion that the precepts of the Sermon on the Mount could be turned back to front in wartime. A subsequent Zeppelin raid on the ironworks at Carlin How did not change his mind.

By the following spring the government knew that the war was going to take many more men than voluntary enlistment could provide (Fig. 5). Legislation was accordingly enacted to discover how many men were available for service and what they were currently doing. The Military Service Act followed early in 1916. The Act deemed all unmarried men between the ages of nineteen and forty-one to be already enlisted unless they had been exempted.[6] Men who objected on grounds of conscience might be freed from combatant service, but not from service itself.

The Act took effect on 2 March 1916. Myers applied for exemption on grounds of conscience and his case was heard at a local tribunal a fortnight later. Myers told the tribunal's four members that he believed all races of man to be one. This being so he could not kill or assist in killing, as it would be a crime to do so. Myers added that it would have been better for humanity if all the churches had spoken out against the war, with which he would not help, even in ambulance work. It was against his principles to take a military oath.[7] The tribunal acknowledged Myers's genuineness and exempted him from fighting, directing him instead to the Non-Combatant Corps – a body under military discipline that did work in support of the war effort. Myers went to the North Riding Appeal Tribunal and got the same result.[8]

No. 2 Company of the Northern Non-Combatant Corps (NCC) was based at Richmond, and it was here that Myers was ordered to report at

5. *Leeds Pals at Colsterdale, near Ripon, October 1914. At this stage in the war social cohesion and public interest in such units was strong; the photograph was one of a series, taken tent by tent, for use as postcards and to keep contact with families and friends. Standing right is Harry Cockram, met later (see Fig. 20).*

the beginning of May. Richmond is a hillside town of cobbled streets, stone-flagged roofs and steep-sloping wynds around a wide market-place (Plate 3). Next to the marketplace is a castle poised on a cliff, at the foot of which flows the Swale, an energetic, glittering river that has carved its way down a dale silted by names from different languages. On its way to Richmond the Swale tumbles over Wain Wath Force, flows through Muker, under Scabba Wath Bridge, past Oldfield Gutter. Below Applegarth parts of the dale are wooded. By midsummer the trees wear a general green, but in late April and early May each kind has its own hue. Tips of birches just before bud-burst are purple. As beech canopies unfurl they form a green-silver mist. Best of all are the oaks, which for a few days are tinged by old gold and bronze.

None of that is likely to have been high in Alf Myers's thoughts on 1 May 1916, although as a devout Methodist he might have reflected on the irony of the Swale's reputation as England's Jordan. Michael Drayton's song about Yorkshire recorded how in the seventh century:

> Paulinus of old Yorke, the zealous bishop then,
> In Swales abundant streame Christened ten thousand men,

> With women and their babes, a number more beside
> Upon one happy day . . .[9]

Obedient to the same faith Myers ignored the direction to report to the NCC. A policeman duly arrived on the doorstep at 1 Steavenson Street and arrested him. Magistrates ordered that he be handed over to his unit at Richmond. On arrival Myers refused to accept orders or wear uniform. He was accordingly imprisoned in Richmond Castle. With him were fifteen other COs, most of whom were members of the Society of Friends, Methodists and the International Bible Studies Association. Their writings and manifestos can still be seen on the cell walls (Fig. 6).

A number of northern MPs got wind of what was happening. On 10 May one after another of them asked Harold Tennant, Under-Secretary of State at the War Office, about reports of ill-treatment, beatings, deprivation of food, public humiliation and a mock execution. After the fifth question Mr Tennant replied:

> 'I am going to make an appeal to my Honourable Friends, and to Members in all quarters of the House, not to press me for answers to these and similar questions involving inquiries into the cases of individuals. The labour involved in procuring answers to such inquiries is enormous. No such staff is available at present, and, if the House considers that answers to inquiries of this kind should be secured, a special staff will become necessary. I am reluctant any further to ask officials, not only in the War Office, but in the various commands, to undertake work of this kind in addition to that with which they are already overburdened.'[10]

Tennant's plea simply prompted more questions.

> Philip Snowden (Blackburn): 'Are these men then to continue to be tortured because inquiry may involve a little trouble at the War Office?'
>
> James Hogge (Edinburgh East): 'Are we to understand that the cases are so very numerous that they require a special staff?'
>
> Robert Outhwaite (Hanley): 'Will the Right Honourable Gentleman say what steps a Member can take when he has brought before his notice what he believes to be cases of illegal and very brutal treatment? Are they to be disregarded because of waste of time?'

And so on. But it was to no avail. On 29 May the sixteen were put onto

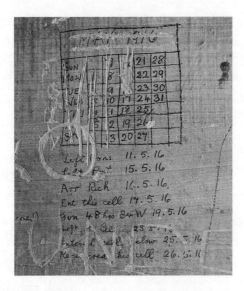

*6. Calendar drawn on Richmond Castle cell wall by John Hubert
Brocklesby, May 1916. Hymns, songs and devotional reflections were also
written on the walls.*

a train at Darlington, taken to Southampton, handcuffed and transferred to the hold of an ageing pleasure cruiser which now worked as
a troopship. With them were COs who had been brought from three
other areas, one of whom recalled that it seemed as though they had
been chosen for the purpose. As indeed they had: their journey was a
contrivance by Field Marshal Lord Kitchener, Secretary of State for
War, to bring a token group of absolutist objectors into the presence of
the enemy, and thereby put them on active service.

They arrived in Le Havre the next day. From here there was a slow
rail journey to Boulogne. The train often stopped, the mood was relaxed;
for hours at a time the men sat in the open doorways of box cars, legs
dangling, enjoying the heat. A guard who stepped down to stretch his
legs absently passed his rifle to a CO and asked for a hand up.

The idyll did not last. At Boulogne they were told that the previous
batch of COs had been shot. This was a ruse to cow them into compli-
ance. It did not work. On 6 June they were ordered to unload supplies
at Boulogne docks, and refused. The order had been given by an officer,
and since the men were now technically 'in the presence of the enemy'
the penalty for non-compliance was death. The Yorkshiremen were put
into the Field Punishment Barracks at Harfleur where they were held
in conditions designed further to break their will.

The prisoners were visited by Frederick Meyer, the Baptist minister of Christ Church, Southwark, and Hubert Peet, a Quaker Socialist. The meeting was held in the camp guardroom under supervision. An account published shortly afterwards in the journal of the No-Conscription Fellowship is poignant: one of the visitors was taking notes, apparently at speed: 'Carlin' appears with a question mark in the list of the men's names.[11]

On 12 June the sixteen were formally charged and convicted at a field general court-martial. At this point one of them yielded. The others were brought to Henriville Camp for sentencing on Saturday, 24 June. The ceremony was contrived in a way to intimidate those who saw it, and to break the spirit of its victims. An audience of several hundred NCC and labour battalion soldiers was drawn up in ranks on three sides of a square. On the fourth side was a platform onto which the prisoners were led. The presiding officer addressed each man in turn, reading out his name, stating that that he had been found guilty, and that his penalty was 'To suffer death by being shot.' Silence followed, eventually broken when the court said that the sentence had been confirmed by the Commander-in-Chief. Another long pause enabled everyone to reflect that this meant there could be no appeal. At length, the officer added that the sentence had been commuted to ten years of penal servitude. The fifteen were not entirely surprised: another group of COs had been similarly dealt with a week before, and on return to barracks had described what had happened.[12] The Yorkshiremen had at first been incredulous.

In early July, Myers was returned to England and put in Winchester prison pending a decision on what to do with absolutist COs. Herbert Asquith, the prime minister, instructed that every case of a court-martialled CO should be reviewed.[13] Cases were to be assigned to one of a number of categories; Category A recognised a genuine objection to all forms of service. Category A objectors were to be released from custody and handed into the charge of the Brace Committee (so called because it was chaired by William Brace MP, Under-Secretary of State at the Home Office), which had been set up a few weeks before to coordinate the employment of COs. Myers was interviewed by the Central Tribunal at Wormwood Scrubs. The tribunal placed him Category A and referred him to the Brace Committee, which in turn assigned him to Dyce, a work camp at a granite quarry on the outskirts of Aberdeen.[14]

When Myers arrived at Dyce at the end of August about 250 men were already there, many of them enfeebled by months of abuse, meagre nutrition and lack of exercise. A large number were from academic or clerical backgrounds with no experience of heavy manual labour, incapable of pushing barrowfuls of granite rubble through gluey mud. At first each man worked for five hours a day, but the Brace Committee thought this too easy and increased the hours to ten. One visitor said the labourers looked as if they had been sleeping rough for twenty years. They slept on straw mattresses in ragged bell tents that had been discarded by the army. Camp organisation was chaotic. The mattresses rested on mud, the men's own efforts to improvise better accommodation in nearby derelict buildings having been vetoed. In the space of four weeks a local doctor certified seventy-three men as ill or unfit for the work. On 8 September one of them, Walter Roberts, died of pneumonia.

The inmates produced a four-page newsletter called the *Granite Echo*. It was edited at Dyce by the anarchist-communist Guy Aldred and passed to London, where it was printed on Aldred's Bakunin Press and distributed through radical circles. The newsletter described camp conditions and printed a charge made by members of the COs' committee: 'We view the present position as extremely serious, and assert that but for indifference and neglect, our late comrade Roberts would be alive now.'[15] Copies reached the No-Conscription Fellowship, which protested to the Home Office.[16]

Ramsay MacDonald, leader of the Labour Party, came to see for himself. Addressing fellow MPs a month later he described the 'swaying masses of mud' and the chaotic management. MacDonald told the House that it was not simply the hopeless state of the men and squalid conditions that had appalled him but also the pointlessness of what the men were being forced to do:

'It was really a most melancholy spectacle. You felt when you saw it. Here is this country at death grips with an enemy, fighting for its existence in a way it has never had to do before, and it ought to make every one of us bend our backs to do the work we can do, and demand that we shall have our opportunities of doing it given us by those who have the power and the authority to assign us that work. There were these men, about a hundred, doing work they were not trained to do, doing work they could not do, doing work they could not be trained to do, going on under the impression that this is national service.'[17]

In reply, William Brace downplayed the significance of the poor conditions.

'But the House will please remember that the Dyce Camp is not of a permanent character. It was simply a camp which was started for two months. It is being closed, and within a week from now the camp will be entirely disbanded.'[18]

Two days later relatives and parents of the COs who had been sentenced to death wrote to the prime minister, Lloyd George, and the Home Secretary in protest at the continued persecution.[19] Later that week Dyce was closed. Myers was granted a week's leave – the first contact with his family since April – and ordered to report to the Army Reserve's Work Centre at Wakefield, in the West Riding of Yorkshire, on 1 November.

After Dyce the regime at Wakefield at first seemed humane and organised. The Centre was in Wakefield prison but it was run by a doctor, locks had been removed, there was communal dining, and the warders wore civilian clothes. The work was basic, but it ended at 5 p.m., whereafter men were free to go out until mid-evening. Wakefield's Quaker meeting house became a social centre, and COs were welcomed by Friends in nearby Leeds and Bradford. Initially they came and went without any public ado, but after a time the local press incited protests against COs, some of whom were beaten up as they returned to their quarters.

In December the Brace Committee directed Myers to work for a private employer. He refused. We do not know whether this was because of the nature of the work or out of objection to releasing another worker for military service, but either way the Brace Committee washed its hands of him and he was returned to prison. For a few weeks he was held in the North Riding gaol at Northallerton; then in February 1917 he was transferred to Maidstone, one of the oldest prisons in Britain. With him were others who had been condemned to death in 1916, like Norman Gaudie, sometime centre forward for Sunderland football club; John Brocklesby, a schoolmaster and Methodist lay preacher from Conisbrough; Fred Murfin, a printer and chapel-goer who eventually joined the Society of Friends. Also with them for a time were other kinds of dissident, like Eamon de Valera and members of Sinn Fein who had been imprisoned after the Easter Rising.

Newcomers at Maidstone underwent introductory hardships and deprivations that were intended to crush their spirits and make them obedient. For the first fortnight a prisoner slept on a plank. He received reduced rations during his first month, and was forbidden to receive or write a letter for the first two. Thereafter he was allowed to receive and write one letter per month. Only short letters were permitted. Long letters were liable to be returned undelivered, and letters out were not to discuss prison life or national subjects. After the probationary period well-behaved prisoners were allowed one thirty-minute visit per month. The Independent Labour Party contributed to a fund for dependants of COs, but the round trip from Carlin How to Maidstone is 580 miles, so Myers received few visits. In September 1917 his mother died.

Prisoners were locked in their cells from late afternoon until 6 a.m. the next day. The cells at Maidstone tended towards climatic extremes, some being cold, others broiling. Not all cells had windows; in those that lacked them a prisoner might stand by the door to read by light borrowed through the Judas hole. Allowable books were a Bible and prayer book together with one volume of non-fiction and one other from the library each month. Petty rules were legion: Myers was not to lie on his bed before bed-time, talk to others for most of the day, lend his library book, look out of the window or own a pencil. In such instances rights to daily exercise, work in association or letters could be taken away. If letters were stopped there was no explanation to the family, whose members were left to wonder what had happened. Bad food and insufficient outdoor exercise led to lowered immunity, with the result that normally minor ailments like colds could be fatal. Early in 1919 two COs in Maidstone died. One of them was Fred Wilkinson, a Christian organiser who caught flu over Christmas and succumbed to pneumonia on 3 January. He was twenty-seven.[20]

Resourceful prisoners kept their minds active as best they could. Cell-to-cell chess, played without a board or pieces, was one pursuit. Moves were whispered at moments of opportunity – during exercise, work in association, or when orderlies distributed fresh underclothing on Saturday afternoons. Games could last for weeks. The event most enjoyed by prisoners was weekly choir practice, when they could give voice together.

So the days passed.

After the Armistice imprisoned COs became restive. Disturbances occurred in prisons across the country. COs in Wandsworth went on hunger strike, broke furniture and engaged in systematic disobedience that continued even after those involved had been put into solitary confinement.[21] On 3 April 1919 Charles Cripps, Lord Parmoor – lawyer, lay churchman and peace campaigner – rose in the House of Lords to call attention to the position of conscientious objectors. He reminded their lordships of a memorandum that had been submitted to the prime minister on 1 January.

'It stated that there were 1,500 men then in prison, 700 of whom had served terms of two years, while pointing out that two years is the maximum punishment allowed for ordinary criminals under either the civil or military code. It pointed out further that a special Inquiry had been made, that it was a laborious matter, and that as a result it had been ascertained that the large majority of these men, who were imprisoned for a period beyond that applicable to the most mean and despicable criminals, were men who had acted under the demands of their conscience and in accordance with deep moral or religious convictions. That statement was made on undoubted authority. The Memorandum was signed by members of your Lordships' House and others whom I have indicated, and yet no answer was made, either officially or unofficially.'

Lord Parmoor continued:

'Time passed along, and when I put down the Resolution the statistics were – they may have altered since – that there were 773 conscientious objectors in prison believed to be perfectly sincere, who had been there for more than two years, that 218 had been court-martialled since the Armistice, some for the third and fourth times, that fifty-nine men had died subsequently to their arrest, and I have noticed a death or two mentioned in the papers since that date – I do not say they all died in prison – and that thirty-nine men had become mentally afflicted.'[22]

On the same day, Winston Churchill, Secretary of State for War, announced an amnesty for most military prisoners.[23]

Alfred Myers was released a week later, on Saturday, 12 April. With him were several other members of the original Richmond Sixteen. One of them, John Brocklesby, recalled the moment. First, a yelled order: 'Brocklesby, Gaudie, Myers, Murfin, wanted!' The men were taken to their cells to collect possessions, then marched through A Hall, where other prisoners were standing by cell doors. As the COs passed there were murmured beginnings of a low cheer. The cheer grew to a roar. It was the day before Palm Sunday. Coincidentally 12 April was also the anniversary of the baptism in AD 627 of Edwin, who ruled Yorkshire's ancestor province of Deira from c. 616 to 633, and his grandniece Hild. We recall that Hild became the founding abbess of the monastery at Whitby, and nemesis of snakes, who were turned into ammonites.

A common experience of returning COs was being spurned by neigh-bours and shunned by employers. Carlin How had lost twenty-one of its men and the fact that Myers was still alive because he had refused to go with them may have made for a difficult homecoming. This said, Myers had been well respected at the mine, and only those who have not worked underground could suppose a miner to be lacking in cour-age. John Hubert Brocklesby, another CO who had been with him at Maidstone, gives us a glimpse of him on the way home: he recalled that 'poor old Alfred' showed signs of distress, and that when the time came for them to go their separate ways Myers felt unable to continue alone.[24]

We can retrace the journey. On release Myers was given a rail war-rant, and next morning took a train north from King's Cross. The Great Northern Railway's main line crossed into Yorkshire at Bawtry, a town planted in the late twelfth century between the Roman road to Doncas-ter and the River Idle at its highest point of navigation.[25] The position permitted Bawtry's emergence as an inland port. The Tudor antiquary John Leland thought the place 'bare and poore', but the combination of water and road links led to a period of affluence in the later eight-eenth century and early nineteenth, reflected in elegant houses, chapels and coaching inns. The Idle is a very modest-looking waterway, but, as Daniel Defoe found in 1724, it was deep enough to float barges for seven miles across to the River Trent, and from there 'quite to Hull'.

> By this navigation, this town of Bautry becomes the centre of all the exportation of this part of the country, especially for heavy goods, which they bring down hither from all the adjacent countries, such as lead, from the lead mines and smelting-houses in Derbyshire, wrought iron and edge-tools, of all sorts, from the forges at Sheffield, and from the country call'd Hallamshire, being adjacent to the towns of Sheffield and Rotherham ... Also millstones and grindstones, in very great quantities, are brought down and shipped off here, and so carry'd by sea to Hull, and to London, and even to Holland also. This makes Bautry Wharf be famous all over the south part of the West Riding of Yorkshire, for it is the place whither all their heavy goods are carried, to be embarked and shipped off.[26]

In the early 1800s Bawtry's reputation as a gateway was boosted by growth in road traffic. This dwindled when the railway came, then resumed with the spread of car ownership.[27] For about forty years a motorist's first experience of Yorkshire on the Great North Road was Bawtry's spacious Georgian High Street. Then a bypass took the motorists away and the association with arrival went with them.[28] The station has gone too, so today's rail passenger finds no ready sign where Yorkshire begins or ends, except sometimes in winter when the Idle is in flood and the railway runs beside a huge lake. This is in contrast to the west coast main line, where Crewe junction was glorified by W. H. Auden as 'the wildly exciting frontier where the alien South ends and the North, my world, begins'.[29]

Ideas about where 'the north' begins, what it represents, how it differs from the south and where the divisions run have been shaped by images, stories and literary traditions that go back for centuries. To some of them we shall come.[30] Meanwhile, if Yorkshire has a counterpart to Crewe it is arguably a bit further on, at Doncaster, where Bert Brocklesby left the train to change for his home town of Conisbrough. 'Donny' is another inland port on the western branch of the Roman Ermine Street, and for the few minutes it takes to reach it the train today passes plantations, sandy lands, flailed hedgerows, gravel pits and carrland.[31]

The Great Northern Railway's Official Guide of 1861 described Doncaster as the 'first town of any importance on the approach from the adjoining counties of Nottingham and Lincoln', and 'one of the cleanest and best built towns in the kingdom'.[32] It does not feel like that today. The streets fronted by Georgian and Regency buildings that impressed earlier visitors have gone, new roads have been hacked through and former industries have been supplanted by titanic, featureless distribution sheds that seem to have fallen from the sky at places where main roads meet.[33] Yet behind the post-industrial mess is a district that in the eighteenth and nineteenth centuries was among the wealthiest in Britain. Good communications, fine building stone, and terrain suited to the arts of landscape architects were among the reasons why Doncaster came to be bordered by ancestral seats, parklands and great houses. Bankers, viscounts and judges lived nearby, and if Myers had stepped off the train for an hour or two that Sunday their affluence would have been evident in things like John Carr's classical grandstand at the race course, the gilded Mansion House, or the soaring parish church of St George that many took for a cathedral.

Myers left the train at York, where the remaining part of his journey can still be traced on a wall map that shows the North Eastern Railway's passenger system as it stood in 1905 (Fig. 7). The map is formed of sixty-four eight-by-eight-inch ceramic tiles. The unique design on each tile was engraved on a copper plate that was used to print a transfer which

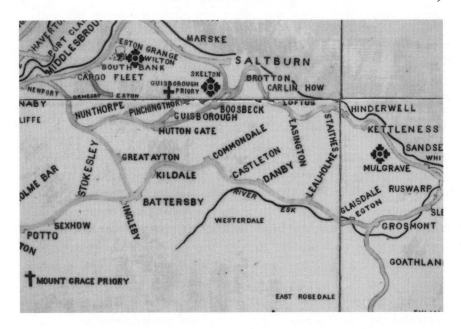

7. Extract of tile map of North Eastern Railway, York station, 1905–10

was placed on the tile, followed by glazing and firing. The result uses three colours on a white background and includes features like castles, abbeys or battlefields that a traveller might wish to see. Around twenty-five such maps were produced, each for a different station. Nine survive at their original locations.[34] One of them is at Middlesbrough, to which Myers now took train.

Along the way, a mile or two beyond Northallerton, Myers could look out at the abrupt scarp of the Cleveland Hills. Cleveland means something like 'the cliff land'. A traveller's first sight of it when approaching from the north – from Edinburgh, say, or Newcastle – is stirring: as you arrive at the lip of the Tees valley Middlesbrough is spread out in the foreground, backed by the hills' undulant frontage that runs eastward from the edge of the Vale of Mowbray for miles until it is cut by the sea. The sense of a beginning here is sharp, quite different from Bawtry's are-we-there-yet borderland. Conspicuous against the Cleveland façade is the crooked profile of Roseberry Topping, a sandstone outlier of the North York Moors. 'Roseberry', from *Othenesberg*, 'Óðinn's hill', is the kind of place that Ælfric, tenth-century abbot of Eynsham, might have had in mind when he preached a sermon against making offerings to pagan gods in high places. Like most of Cleveland the hill has been tunnelled and pitted by miners; in 1913 its rose-thorn outline was sharpened by a huge rockfall. Roseberry Topping rises above 1,000 feet and those who live within sight of it say that when the summit is in cloud rough weather will follow. This was the case that April Sunday, when Yorkshire was under grey overcast and a depression was heading in from the Atlantic.[35]

At Middlesbrough Myers changed again, now for Saltburn-by-the-Sea where he had to wait before the last leg of his journey. If he sat on a station bench or walked the platform, listening to strident gulls, he could have looked down the track that continued to a private canopied dais adjoining the Zetland Hotel, where first-class passengers could step out of their coach and walk straight into the hotel. The Zetland took its name from Lawrence Dundas, the third Earl of Zetland, on whose land much of Saltburn was built. The man who had invited designs and estimates for this 'commodious hotel' back in 1860 was Henry Pease, a railway promoter whose family owned collieries, ironstone mines in Cleveland and lead mines in the north Pennines.

Founding things is what members of the Pease family did. Joseph, pioneer of ironstone mining at Skinningrove, was described by *The Times* as the 'founder of Middlesbrough'. Henry himself had founded Saltburn, 'the watering place at the terminus of the Darlington section of the North Eastern Railway'.[36] Their father, Edward Pease, had been a leading backer of the Stockton to Darlington railway in the 1820s. The late Victorian public regarded Edward as the founder of railways.

It was railways that made Saltburn possible. Aside from some fishermen's cottages where the valley of the Saltburn Beck met the coast the place had no practical existence until the Pease family took its fortunes in hand.[37] If Myers had stepped out through the station's classical portico and gone for a walk he might have admired the iron pier, opened in 1869, the faux vernacular kiosks at the pier entrance, or the water-balanced cliff lift by which visitors arriving by steamer could be carried up to the town. Before the war Saltburn had been described as 'a regularly-built town of exceeding charm, with bracing air and lovely views. It is a select place . . . entirely free from the noisy tripper and any undesirable summer visitors.' The admirer continued:

> Strict rules and regulations as to buildings and streets, their number and arrangement, their size and style; strict conditions of tenancy by folk who come new to the town as residents; a close watch on all

undesirables who would affect the good fame or the prosperity of Saltburn . . .[38]

At first sight, then, Saltburn might not be a place in which a returning gaolbird CO would expect to feel welcome. But, there again, Joseph Pease had campaigned for prison reform, and Henry Pease, a Quaker, had been president of the Peace Society. Why it was that so many of Yorkshire's dynamic industrial and social pioneers were Quakers is a subject to which we shall come.

Late in the afternoon Myers boarded the train that would take him home. Carlin How shared a station with Skinningrove on the line to Whitby. In 1871 the prospectus for the line noted that the district formed 'one of the richest portions of the well-known Cleveland iron field, and only awaits opening up by railway communication to equal any other part of the north-east district in mining and smelting operations. It also abounds in building and cement stone, and alum and Petroleum shale'.[39] The prospectus might also have drawn attention to the line's scenic drama. Around 1540 the antiquary John Leland described the coast between Whitby and Teesmouth as a 'cliffy shore'. In places the line ran along the brink of cliffs, gave far views to successive headlands, and crossed gorges and rocky inlets on trestles so slim that from a distance the trains seemed to float in the air. If the line still existed it would probably be a World Heritage Site.[40]

The train set forth: over wooded gills, passing behind terraces of industrial housing, skirting waste heaps. It was no more than ten minutes to the first stop at North Skelton, another five to Brotton. From here the line turned north through coastal farmland to skirt Warsett Hill. At Hunt Cliff the hill is cut by the sea and the track followed the cliff edge for half a mile, giving sight of kittiwakes and gannets. Looking down, Myers could see the wide foreshore platform of mudstone 300 feet below, and had a first glimpse of Cattersty Sands. Then the train turned inland to Carlin How. A clank of couplings, hissing steam, the booming slam of coach-built wooden doors: he was back. As the train shuffled away towards Loftus there were familiar sights to take in: the haze of red dust around the works; Downdinner Hill; hues of buildings made from locally quarried stone, mostly grey and brown, but here and there a gleam of sulphur or reddish chocolate.

And just there, on Skinningrove's single platform, the story fades. History is only as good as its sources, and once the state lost interest in persecuting Alf Myers the sources that tell us about him thin out.

Unlike some of his CO colleagues he wrote no memoir and left no letters or archive about which we know.[41] He did not marry, so there were no children or grandchildren to ask him questions or pass on his answers. How did he fare psychologically? How did members of the chapel react to his return? Did he rejoin the choir, or return to his post as Sunday school superintendent? We have no idea. All we have are the sparse records of civil administration – details like his address, the date of his death, the testing of his will. Like shards from a smashed pane of glass these are bits of the original, but too few and splintery to be reassembled into anything that looks like the whole. However, if we put Myers's fragments alongside those available for members of his family and neighbours, a little more can be said.

Myers's father, Benjamin, was born in 1838 in Dudley Hill, a district of Bradford in the West Riding. His father was a collier and so were his uncles. By thirteen Benjamin was working as a hurrier in one of the collieries that supplied Bowling ironworks with fuel. A hurrier was a child who pulled a coal-filled tub along a roadway. Hurriers were harnessed to the tub by a belt and helped by two more children – 'thrusters' – who pushed from the back. Benjamin's younger brother John was working thus by the age of eleven, and their younger sibling, Thomas, was minding a horse that turned a winding engine at the age of nine. One reason why children were favoured as thrusters and hurriers was the low clearance in many colliery roadways. By the time Benjamin and Thomas were working the minimum legal age was ten, but a report in 1847 noted the temptation to employ boys who were younger: mine managers considered that in seams of coal between eighteen inches and two feet thick, the work was 'done at a disadvantage' unless children were 'brought to it from their earliest years'.[42]

Hurrying was hazardous, but no more so than breathing. The area around Bowling ironworks was known as Bowling Hell. On most days it was overhung by sulphurous fumes and smoke from glowering cinder hills. In its midst was a smouldering heap of clinker that grew annually, and went on burning for years after the works closed.[43] While Bowling was famed for respiratory diseases as well as iron, Bradford at large had gained the reputation of being England's filthiest town. 'If anyone wants to feel how a poor sinner is tormented in Purgatory,' wrote a young German tourist in 1846, 'let him travel to Bradford.'[44] In Bradford the becks ran with cess. Cholera and typhoid prospered. Buildings and trees were furred with soot. Thousands lived in sodden hovels, sloshing to work through stagnant pools in which bobbed human turds, along streets in which violence and disorder were rife. At the time of Benjamin Myers's birth average life expectancy in Bradford was just over eighteen years.

On Christmas Eve 1840 the editor of the *Bradford Observer* attempted to stir readers' consciences about health in towns. He quoted from a recent report about nearby Leeds:

> In the Boot and Shoe yard there are a number of rooms, inhabited by three hundred and forty inhabitants; the number of rooms being about fifty-seven, with an average of *six persons to each room*; the annual rent being about two hundred and fourteen pounds. There are *three* out-offices, from *one* of which, during the period of the cholera, *seventy-five* cart-loads of soil were removed by order of the Commissioners, and which is reported not to have been cleaned out since. There is no water within a quarter of a mile; very few inhabitants possess vessels to hold or to attach water.

The editorial continued:

> Such is but a specimen of the dirt, degraded, and wretched state in which the population is frequently found in those districts which our manufactures have crowded with human beings, and for the promotion of whose health and comfort the paternity of government remains to be employed.[45]

One who was trying to exercise such leadership was the Revd Dr William Scoresby (1798–1857), who had been inducted as vicar of Bradford the year before. Before entering the Church, Scoresby had had an earlier career as a master whaler in the high Arctic.[46] He arrived in Bradford in a spirit of reforming zeal. He campaigned for cleaner air, better housing, and local schools for children of colliers and millworkers.[47] After eight years he left, disheartened by lack of support for reform and worn down by hostility to his plans for pastoral reorganisation. In 1840 the parish of Bradford was much the same as it had been in 1140 – about fifteen miles across. The parish was originally intended to cater for a small rural population and unfitted to meet the needs of the ever-expanding, raucous, restless industrial conurbation that it now contained. Scoresby's plan was to split the area into smaller districts, each with a daughter church serving around 3,000 parishioners. The parish encompassed outlying chapelries whose inhabitants resented the obligation of paying church rates both to their curate and to the mother parish, while Methodists

8. *Haworth parsonage*

and other dissenters begrudged paying anything. Incumbents of several chapelries were opposed to Scoresby's plans to divide them on the grounds that this would reduce their own income. Bradford's westernmost chapelry centred on the township of Haworth and contained 6,000 people, many of whom wished to separate from Bradford. This posed a dilemma for Scoresby, as Haworth contributed a fifth of his income. Haworth's parson was a tall, formidable Evangelical, often to be seen striding about among his parishioners or across moorland to distant farms (Fig. 8).[48] His name was Patrick Brontë.

It was not just higher wages, then, but hopes for a better life that led Benjamin Myers to move his growing family out of polluted, life-threatening Bradford, first to the newly opened ironstone mines at Eston, on the edge of Middlesbrough, then to Skinningrove. Census records enable us to follow them. In 1881 Benjamin (now forty-six) and his wife, Harriet, were living in nearby Loftus. Harriet was the daughter of an agricultural labourer in the East Riding, and by now had borne six children, the last five at two-year intervals. By 1891 Benjamin was dead and Harriet was head of the family. Two of her sons, George (fifteen) and William (twenty-two) were working in the ironstone mine, and a third, Tom (eighteen), had been doing so until a mining injury left him disabled. At this point Alfred was six, the last but one of ten children, one of whom had died. The family now lived at 3 Wood Cottages in Carlin How. The Wood Cottages were three rows of makeshift housing erected by Bell Brothers, owners of the mine, to accommodate workers who had been brought in from outside. They were small – three rooms – and so called because they were made of wood. Residents shared one tap and used dry privies, but there was space for allotments on the hillside and many families kept a pig. Although intended as temporary the cottages were still occupied fifty years later, by which time they were remembered as Bell's Huts.

Bell Brothers also established several brick-built terraces of two-up, two-down houses up on the plateau at Carlin How, and it was into one of these in Steavenson Street that Alfred Myers and his mother moved in the 1900s. Thomas, meanwhile, had recovered sufficiently from his disability to return to work in the mine, where he was now a deputy. The 1911 census finds him and his young family living in another of the Bell's Huts, while Alfred's younger brother Henry had returned to the West Riding to work as a collier. Like his father's, Henry's mining career began early: by the age of fourteen he was working as a trapper boy. Trappers were the youngest to work in Bell's Pit. They did so in darkness, operating doors that governed airflow and ventilation in different parts of the mine, opening the traps to enable horse-drawn tubs of ironstone to pass and closing them as soon as they had gone through.

All Benjamin Myers's sons worked as miners and back in the West Riding all his brothers and nephews did so too. Curiously, the same names recur in each branch of the family and in successive generations.[49] Typically, the boys started in a mine as children or young teenagers. The dangers and hardships they faced were sustained and extreme, but mine work was better paid than agricultural labouring and the presence of three or four wage-earners in the Myers household meant that family income was sufficient for more than daily needs; infants and youngsters could be cherished and family members who were too old to work could be kept out of the workhouse and cared for at home. Family solidarity extended to the mine, where the practice of working in pairs, a miner and a filler, the miner being the more experienced partner, was suited to a father and son or elder and younger brothers.

We cannot know how Benjamin Myers or his children saw themselves, but there are signs that their sense of identity centred less on where they were than on what they did and the relatives and neighbours with whom they did it. Such communities based on family and occupation linked readily with other groups formed through pastimes and shared pursuits. At Carlin How these included brass bands, choirs, an allotment society, pigeon racing, cycling, fishing and sword dancing. From 1866 the railway enabled those who took part in such activities to join with others through day trips and outings in doing so.[50]

Did such networks foster a sense of Yorkshireness? Yorkshire as a political concept became blurred during the nineteenth century, first by the Reform Act of 1832, which replaced the county constituency centred on York with parliamentary divisions for each Riding, and then from 1888 by the Local Government Act, which turned each of the three Ridings into an administrative county in its own right.[51] This concluded a process begun at the Restoration whereby each Riding acquired its own personality, reflected in separate quarter sessions, militias, lieutenancies, and occasions like race meetings, performances, hunts and social seasons that enabled normally dispersed gentry to come together. With this went the gradual displacement of York by the emergence of county towns for each of the Ridings: Wakefield, Northallerton and Beverley.[52] Alf Myers's sense of Yorkshireness was very likely coloured by the time he had spent in the prisons at two of these three places.

The years when the Myers family mined coal under Bradford and tunnelled for ironstone under Cleveland coincided with mounting interest in the historical Yorkshire, its people and landscape. Among groups formed to advance inquiry into different aspects were the Yorkshire Philosophical Society (1822), the Yorkshire Geological Society (1837), the Yorkshire Archaeological Society (1863) and the Yorkshire Dialect Society (1897). These were pioneering steps: the Yorkshire Geological Society was only the fourth geoscientific body to be formed in Britain, while the Dialect Society was the first of its kind in the world. In practice, however, most of these organisations sprang not from Yorkshire at large but from the industrial West Riding, where mining and coal provided a thriving context for geological interests: the Archaeological Society grew out of an association in Huddersfield, and the Dialect Society evolved out of a committee that met in Bradford to supply information for a national dictionary.[53] The Yorkshire County Cricket Club was formed in Sheffield in 1863, but for a time it was only one of several clubs that claimed to represent Yorkshire, and at least one of its rivals was likewise based in the West Riding.[54]

9. *Poems, plays and novels connected with Yorkshire*

There are different ways to come to Yorkshire – in a poem, through an image, on a ferry from Rotterdam, by train. The main channels through which ideas of 'Yorkshireness' spread during the later nineteenth century and early twentieth were popular print media. From the 1860s references to different aspects of Yorkshire culture abounded in newspapers, periodicals, illustrated ephemera, almanacs, postcards, cartoons, ditties, fiction.[55] Guides for ramblers formed one widely circulating genre. For reasons given there was widespread eagerness for getting out of cities, even if only for a few hours at a time. In the later nineteenth century regional papers like the *Leeds Mercury* opened 'doors of delight' to 'tired city workers' by publishing weekly rambles 'for all who heed the call to the greenwood'. Periodically these were gathered up and issued as inexpensive free-standing booklets, which in their turn were revised and reissued in successive editions.[56] The arrival of halftone reproduction in the 1880s enabled the illustration of such guides with photographs of Yorkshire beauty spots, landscapes and famous buildings.

On a larger scale were compilations of stories, places and landscapes in volumes written for the general public. Fred Cobley's *Upper and Lower Wharfedale* (1890), Harry Speight's *Nidderdale and the Garden of the Nidd, a Yorkshire Rhineland* (1894) and Edmund Bogg's *A Thousand Miles in Wharfedale* (1892) give the flavour. Speight (1855–1915) grew up in Bradford, where he worked in a dye works run by his father. Material for his books on Nidderdale, Craven, Wharfedale, Airedale and Richmond was collected in the course of cycling and walking tours (which in the 1890s took him out of his home in smoky Bowling), combined with library work, antiquarian collecting and interviews. Edmund Bogg (1850–1931) was a dedicated rambler who ran an artists' materials business in Leeds. His output was torrential, became increasingly slipshod and was latterly recycled 'using poorer quality paper and paperback binding'.[57] It mattered not: 'these later, cheaper, editions brought the countryside to a less affluent and increasingly mobile working class population'.[58] Similarly populist was the Halifax-born Joseph Fletcher (1860–1933), a journalist who learned his craft in London and returned to the West Riding to work on the *Leeds Mercury* and *Yorkshire Post*. His productivity included Yorkshire novels, history and historical fiction, and, from 1914, detective stories.

Probably the most widely read Yorkshire novel of the early twentieth century was *Windyridge*, the story of an artist-photographer who moves from hectic London to sublime Yorkshire. Its author was William Riley (1866–1961), a Bradford businessman who wrote it during 1911 as a diversion to be read aloud in instalments to his wife and some friends. The hearers encouraged Riley to offer it for publication. Herbert Jenkins, a new publisher in London, accepted it and within four years it had sold 150,000 copies. When Riley died in his ninety-fifth year sales had topped half a million.

Windyridge became 'the name of countless semi-detached houses owned by expatriate Yorkshiremen'.[59] It is worth pausing to look into the background from which this sentimentality came. Like Harry Speight, Riley started out working for his father. Like Benjamin Myers, and close to him in time and place, Joseph Riley's upbringing had been harsh. He was born in 1838 into a poor Bradford family, the fifth of eight children, and sent to work with a rope-maker at the age of seven. Like Alfred Myers, he became devout, and membership of a Methodist congregation changed his direction. Joseph was encouraged to attend evening classes, and after a succession of different jobs he and his brother Sam went into business on their own account in the stuff trade.[60] In 1883, Joseph Riley bought a magic lantern for his sons William and Herbert. Riley saw a commercial opening for associated products and set up a lantern-slide business. Within ten years the enterprise was trading internationally. The Rileys progressed to projectors for the new 'living pictures', and from there to making their own short films. The slide and projection business continued under William Riley's direction until 1914, by which time his new career as a writer had been launched.[61]

The success of *Windyridge* was helped by the marketing skills of its publisher, Herbert Jenkins. Jenkins had a keen eye for covers and straplines:

> Copies of *Windyridge* were sent to Yorkshire booksellers with a special cover band stating 'The Great Yorkshire Novel'. Those sent to strongly Methodist towns had a cover band stating 'The Great Methodist Novel'. When a reviewer compared *Windyridge* to *Cranford*, Jenkins began marketing the book in literary circles as 'The New Cranford'.[62]

Aback of this were ways in which Methodist culture and Yorkshire's sense of itself interacted. Methodism spoke to working-class communities in words they could understand and encouraged people with little – like Willie Riley's father and Alf Myers – to advance through education. Both physically and figuratively chapels were at the hearts of places: they brought people together socially as well as for services, and both types of occasion gave opportunities for shared enthusiasm for music and communal singing. Methodism transcended class; it

became a movement with which many Yorkshire people identified, and
its plain-spokenness provided a model for new writers.

Alongside the West Riding authors were antiquaries and topog-
raphers who specialised thematically or pioneered the study of their
own sub-regions. Cleveland's relator was the antiquary and poet John
Walker Ord (1811–1853), whose interest in aristocratic breeding imparted
a special fondness for the neighbourhood of the Skelton Beck. ('From
this little nook of Cleveland,' he wrote in 1846, 'sprang mighty mon-
archs, queens, high-chancellors, archbishops, earls, barons, ambassadors,
and knights; and above all, one brilliant and immortal name – Robert
Bruce.'[63]) Richard Vickerman Taylor (1830–1914), a Leeds schoolmaster
and cleric, collected Yorkshire biographical anecdotes and wrote about
churches.[64] John Mortimer (1852–1911), a corn chandler in Driffield,
made ground-breaking landscape studies of the Yorkshire Wolds. Such
eagerness carried into the twentieth century. The regional monthly
the *Dalesman* was founded in 1939 and has flourished ever since. The
magazine's enthusiasm for Yorkshire stories, humour and dialect was
prefigured in the region's newspapers at least from the later nineteenth
century.[65] During the 1930s and 1940s Marie Hartley (1905–2006) and
Ella Pontefract (1896–1945) celebrated the individuality of different
Dales, collecting material that illustrated traditions and lives in days
when things like cheese, shoes or ironwork were still made locally. After
Ella's death Hartley joined with Joan Ingilby to produce over twenty
books, at least two of which – *The Old Hand Knitters of the Dales* (1951)
and *Life and Traditions in the Yorkshire Dales* (1968) – became local-
history landmarks. Hartley, Ingleby and Pontefract were acquainted
with Arthur Raistrick (1896–1991), who grew up in the industrial
village of Saltaire outside Bradford and left school at sixteen to be ap-
prenticed as an electrical engineer. Like Alfred Myers he joined the
Independent Labour Party and became an absolutist conscientious
objector, for which he was imprisoned. On release in 1919 he joined the
Society of Friends and embarked on a career embracing geology, industrial archaeology and the landscape history of the Yorkshire Dales.[66]
With all this attention, then, small wonder that two out of the first
ten national parks to be designated in England and Wales should be
in Yorkshire.[67]

A key point about this output is that while its subject was Yorkshire at large, by far the largest part of its Yorkshire audience was in the
industrial West Riding, and much of the rest lived outside Yorkshire.
In 1831, 72 per cent of Yorkshire's 1.33 million people lived in the West

Riding. By the century's end the respective figures had risen to 77 per cent and nearly 3.6 million. In 1900 more people lived in Leeds than in the entire North Riding, while the combined populations of Sheffield, Bradford and Leeds exceeded those of the North and East Ridings put together. Down to the Great War, then, Yorkshire's sense of individuality in England was fostered by interplay between commercial print media and a mass urban audience.

Alongside this ran an expansion of England's sense of Yorkshire. Such awareness arose in a number of ways: through literature, painting, networks of landed gentry and new rich, and latterly through mass travel facilitated by railways. To illustrate: Charlotte Brontë's novel *Jane Eyre* and Emily Brontë's *Wuthering Heights* both appeared in 1847. *Jane Eyre* achieved immediate popular and critical success. Its first printing sold out within weeks and it was reprinted four times in three years. The book was first issued under a pen name, but by early 1850 Charlotte Brontë's authorship was known and literary tourism into the landscapes of her stories had begun. As Charlotte wrote to her friend Ellen Nussey in March that year:

> Various folks are beginning to come to boring Haworth, on the wise errand of seeing the scenery described in 'Jane Eyre' and 'Shirley'; amongst others, Sir J K Shuttleworth and Lady Shuttleworth have persisted in coming: they were here on Friday.[68]

By the time Patrick Brontë died in 1861 annual visitors to Haworth were numbered in thousands. Many of them came from overseas.

Elizabeth Gaskell's *Life of Charlotte Brontë* (1857) stirred fresh interest in the Brontës and Yorkshire. Shortly after its publication Arthur Nicholls (Charlotte's husband) reflected that while Haworth might be a strange place, 'it was nowhere quite as queer as Mrs Gaskell had made out'.[69] Gaskell went out of her way to dramatise the West Riding and its people to contextualise Brontë writing and subject matter.

> Their accost is curt; their accent and tone of speech blunt and harsh. Something of this may, perhaps, be attributed to the freedom of mountain air and of isolated hill-side life, something be derived from their rough Norse ancestry. They have a quick perception of character, and a keen sense of humour. The dwellers among them must be prepared for certain uncomplimentary, though most likely true, observations pithily expressed.[70]

Gaskell reflected that even Lancastrians were 'struck by the peculiar force of character' which Yorkshiremen displayed. 'This makes them interesting as a race; while, at the same time, as individuals, the remarkable degree of self-sufficiency they possess gives them an air of independence rather apt to repel a stranger.' In Gaskell's eyes natives of the West Riding were typified by 'strong sagacity' and 'dogged power of will'. Yorkshire was a place where 'each man relies upon himself, and seeks no help at the hands of his neighbour'. A Yorkshireman's affections were strong, deep-founded, but well concealed.

> Their feelings are not easily roused, but their duration is lasting. Hence there is much close friendship and faithful service; and for a correct exemplification of the form in which the latter frequently appears, I need only refer the reader of Wuthering Heights to the character of 'Joseph'.[71]

The Times gave Gaskell's Life a full-page review, which opened by declaring that there were many reasons why it 'should be read with avidity'.[72] Popular interest in the Brontës and their world was indeed avid, and Gaskell's stereotypes spread accordingly. Or were they stereotypes? In 1968 Phyllis Bentley published an essay entitled 'Yorkshire and the novelist' in which she argued that Yorkshire fiction is characterised by an underlying realism and clearness of diction: a 'determined preoccupation with ordinary people and ordinary lives'.[73]

Smith, Elder & Co was the London-based publisher both of Charlotte Brontë's novels and of Mrs Gaskell's Life. In 1872 the firm began to prepare an illustrated edition of Brontë's work. George Smith and the illustrator took advice from Ellen Nussey and Mrs Gaskell about models for places and buildings. The ensuing depictions prompted a quest for the models of the Brontës' locations. By 1888 'Brontë country' was the title of a book.[74] Fourteen years on, Brontë buildings and landscapes were further investigated by Herbert Wroot in Sources of Charlotte Brontë's Novels: Persons and Places. It has been suggested that the idea of 'Brontë country' did not become popularised until just before the First World War.[75] In fact, the phrase was in general use at least from the earlier 1870s,[76] and by the 1890s it was featuring in advertisements.[77] In 1899 the author of the newly published Highways and Byways in Yorkshire was gently admonished by a Scottish reviewer because he had not mentioned Brontë country.[78]

Settings in Brontë stories continue to fire imaginations. Thornfield Hall in Jane Eyre, for instance, was probably patched together from features in several places. Erskine Stuart identified at least three of them, one being Norton Conyers in the Vale of Mowbray, a house with sixteenth- and seventeenth-century fabric around a medieval core.[79] Its fifteenth-century occupants were the Norton family, old Yorkshire gentry, some of whom fell to the displeasure of Elizabeth I following the Rising of the North.[80] Since the seventeenth century Norton Conyers has been the Yorkshire seat of the Grahams. It is thick with atmosphere, and like many country houses the legend of a mad woman in a remote attic attaches to it.[81] The discovery in 2004 of a blocked staircase leading from the first floor to the attic storey stirred world-wide interest.[82]

A key point about the Brontës' influence on perceptions of Yorkshire is that the British publishers of early editions of their work (as of *Windyridge*) were headquartered in London, and that Smith, Elder & Co – the most successful of them – sold to the world. *Wuthering Heights* (1847) and *The Tenant of Wildfell Hall* (1848) were initially issued by Thomas Cautley Newby, but Smith, Elder quickly acquired the rights to *Wuthering Heights* and published successive editions of Brontë work. George Murray Smith (1824–1901) was a shrewd and effective businessman.[83] An American edition of *Jane Eyre* was in print by 1848. Continental editions of Brontë work quickly followed, and Smith was assiduous in promoting sales across the expanding Empire, notably in India. Thus it was that when Sir Albert Rollitt MP (1842–1922), former mayor of Hull (1883–5), addressed guests at the inaugural dinner of the Society of Yorkshiremen in London in 1891, the Brontës were close to the top of his list of Yorkshire people who were eminent 'in every sphere of action'.[84]

Yorkshire, it has been said, 'is a continent unto itself, each of its three Ridings contributing authors of worldwide fame, but local in their aspiration'.[85] If there is an equal and opposite counterpart to Brontë country it is the low-lying, gently undulating plain of Holderness, Hull's 'remote three-cornered hinterland' between the Wolds and the coast.[86] It hums with names and echoes: Skeffling, Sunk Island, Ravenser Odd. It has soft low cliffs and caravan sites gnawed by a sea that froths brown with memories of places it has already taken.[87] Holderness, too, has had its laureates, notably the novelist Edward C. Booth (1872–1954), and work of Philip Larkin, to which we shall come.[88]

Popular discovery of Yorkshire coincided with the destruction of much that the discoverers wished to celebrate. Bowling Hell was already well alight when the landscapes of Thornfield and Wildfell Hall were put before the world, and depending on the wind the moors around Haworth were often 'dim and lightless' under drifting smoke.[89] Watercolours by landscape painters like Turner's friend Thomas Girtin (1775–1802) and William Callow (1812–1902) on the eve of industrialisation depicted noble monastic ruins clothed in flowers and views of places like Richmond that epitomised Yorkshire as it was before it all went wrong. Again, a link between Yorkshire and London was influential. The antiquary James Moore (1762–99) hired young artists to make finished drawings from his sketches for his volumes on monastic remains and castles. One of them was Girtin, and in 1796 Girtin came north to explore on his own account. Over thirty-five watercolours of scenes in dales between the Wharfe and the Tees are the result.[90] Girtin's study of Wetherby Bridge (1800) is an essay in cultural memory and calm. The bridge spans the River Wharfe, bearing the Great North Road on its way from Doncaster up to Boroughbridge and Catterick. Two arches frame buildings beyond. The buildings look a bit shabby – walls are cracked, plaster flakes, and one of the buildings is untidily thatched. Yet windows and mouldings tell of Jacobean structures that were elegant when they were new. Among them is a water mill, its stationary wheel peeping from behind a pier. The river level is low, redolent of summer stillness; reflections from its surface flicker on the soffits of the bridge arches; the drought enables a lady to come right to the waterside, kneeling on tabular bedrock to wash sheets. Apart from the bridge itself, none of this survives.

Turner visited Yorkshire for the first time in 1797, possibly at Girtin's suggestion.[91] Turner himself worked extensively in Yorkshire. He had sympathetic patrons in the Wharfe valley north of Leeds, Edward Lascelles at Harewood and Walter Fawkes of Farnley Hall near Otley. From 1808 Turner was an annual visitor to Farnley, where he came to be looked upon as a member of the family, and his maladroit handling of gigs and carriages in the course of sketching travels earned him the nickname 'Over-Turner'.[92] Turner was fascinated by the weather, scenic spectacle and life of Wharfedale and Washburndale. He was also captivated by Leeds – a city he visited more often than any other outside London, and the manufacturing base of the flax tycoon John Marshall (1765–1845), who bought a number of Turner's paintings. Turner's 1816 panorama of Leeds from Beeston Hill (Plate 4) has been described as possibly the first industrial cityscape in the world.[93] Turner caught the town in rapid transition. David Hill points out that almost all the industrial buildings in his prospect were no more than twenty-five years old.[94] There was as yet a great deal of green. New mills, workshops, runs of industrial housing and busy tradesmen were intercalated with fields and hedgerows, and overlooked by open moorland.[95] Within a generation, dense suburbs had spread up the hillsides; within two, the moors had been built upon and John Atkinson Grimshaw (1836–93) was painting lamp-lit suburban lanes that had taken their place.[96]

Other portrayals of Yorkshire are less known outside their own localities or subjects. Goole-born Reuben Chappell (1870–1940), for instance, emerged as a 'pierhead painter' in the early 1900s, producing pictures of vessels for members of their crews and documenting the busy life of the Humber.[97] Or Thomas Burton (1866–1941), one of two brothers who ran a house-painting business in Beverley, a town which in 1697 had been described by the traveller Celia Fiennes as 'very fine ... for its size', with markets for beasts, corn and fish, and spacious streets fronted by buildings that were 'new and pretty lofty'.[98] After 1700 the town's status as the East Riding's capital led to purpose-built assembly rooms, a theatre, library, a promenade to the north of the town, new houses and the cladding of older structures with Georgian brick.[99] By the 1880s parts of this heritage were in turn being pulled down or altered. Burton's drawings and etchings provide an unequalled record of Beverley's buildings and spaces before they were changed or lost.[100]

Turner, Girtin and Burton all worked on the Yorkshire coast, although as far as we know none of them painted the secluded, wooded valley of the Kilton Beck that rises on the flank of Stranghow Moor and enters the sea at Skinningrove. Someone who might have done so was George Weatherill (1810–90), a local watercolourist whose paintings of Whitby and surrounding places earned him the nickname 'Turner of the North'. Weatherill grew up at Staithes, just down the coast from Carlin How, and must have known the valley when it was still a wooded solitude, perhaps in that week of May when bluebells have 'power upon the soul / To consecrate the spirit and the hour'.[101] Michael Drayton singled out the area as a 'second paradise / Whose Soyle imbroydered is, with so rare sundry Flowers'.[102] In 1875 a correspondent writing to the *Guisborough Exchange* could think of no lovelier place than Skinningrove valley as it had been back then. But now the zig-zag railway had been carved down the hillside, furnace glow bronzed the undersides of clouds, cheap houses and waste heaps cluttered the valley and the beck ran red. 'Why,' asked the writer, 'should such disfigurement and dirt everywhere accompany what we are pleased to call the advance of civilisation?'[103] Whether the Myers family lamented Skinningrove's industrialisation is another question. Like my grandparents, and their friends the Larks, they were glad of the wages from Bell's Pit and the ironworks.

The lives of Alf Myers's family and mine brushed. Myers taught at the Methodist Sunday school attended by my mother and her sisters; during the First World War they sheltered from Zeppelins in tunnels beside Kilton Beck. After that war my mother and her family emigrated. In 1946 she came back. Her home was now in the Midlands, but from time to time there were trips to Carlin How, where Aunt Hattie was her only close relative on this side of the Atlantic. During the first of them Alfred Myers and I came close, for I had just been born and he was about to die.

The earliest visit to Carlin How about which I can remember anything was around 1952. Very probably this is a memory that conflates episodes from different stays. Whenever it was it began on Birmingham's New Street station, which was then a kind of iron-framed cathedral sheltering long curving platforms whence trains smelling of sooty steam and carriage cloth departed for yearningly distant places like Exeter and Carlisle. During the journey my mother pointed to the warped spire of Chesterfield parish church. It was around here, she said, that the Midlands were ending and Yorkshire was about to begin. At Darlington I stood before Locomotion No. 1, early workhorse of the Stockton to Darlington railway, which was mounted on a blackened podium beside the platform where we changed for Saltburn. The final leg was in a single-deck Bedford bus with a backward-leaning pose, its door and windows stylishly aslant.

We stepped off the bus in Carlin How Square, whence it was maybe two minutes' walk to Aunt Hattie's terraced house at No. 2 Queen Street. Queen Street ran parallel to Coronation Street and Gladstone Street, the names together recalling public sentiment in Edwardian days when the terraces were built. Outside, an unending procession of gently nodding buckets floated past on an aerial ropeway that carried ironstone up from the mine. Across the main road stood the blast furnaces, coke ovens, rolling mills and yards of the Skinningrove Works. Before them ran sidings and tracks for mineral wagons and slag bogies. The tracks were screened by a planked fence, but some of the planks had worked loose and a six-year-old could peer through the gaps to follow the to-and-fro journeys of plump tank locomotives with names like Roseberry and Cattersty.[104]

Beyond and below was Skinningrove's shingle beach, where a cast-iron outfall discharged into the sea. Some of the miners and ironworkers worked as fishermen in their spare time, putting out for crabs and lobsters. Their parked boats rested at odd angles beside ramshackle huts in which they kept their gear. Skinningrove is only properly visible from the sea, which perhaps has to do with its local reputation for insularity.[105] Sometimes we walked inland to Skelton Woods, which were deep and shady. My father said there were tigers in the shadows, but I found only primroses.

Mining is a passing business; its communities do not last long. When Hattie's husband, Fred, retired at the end of the 1950s, most of the neighbouring mines were closed, either because they were worked out or because they were being out-competed by open-casting and imported ore. The last of them shut in 1964. At the end of the nineteenth century a journalist speculated that if Middlesbrough continued to develop at the same rate as it had since 1830, it would be among the world's greatest industrial centres by 2000. In the event, that year found most of Middlesbrough's wards to be areas of extreme want, and three of them to be among the ten most deprived in England.[106] Chris Killip's sublimely bleak photobook *In Flagrante* (1988) depicted everyday scenes of deindustrialising communities in north-east England during the 1970s and 1980s.[107] Skinningrove was one of his subjects. At the time of writing the village has 25 per cent more benefit claimants than the national average, three listed buildings, no protected monuments and no tree preservation orders. In Carlin How there are not even that many trees. Nonetheless, it is where our world begins. Rising to the east of Carlin How, between Loftus and the sea, is an undulating plateau with features bearing names like Butter Bank, Gallihow and Downdinner Hill. In Alf Myers's day no one guessed that up here was a secret that would throw light on how Yorkshire began.

Between 2004 and 2007 archaeologists working at Street House Farm between Gallihow and Upton Hill came upon 109 graves. The burials were arranged in rows to make the sides of a square. Inside the square was a central mound, and traces of a lady who had been laid to rest on a sumptuous bed. All traces of the people had been destroyed by the acidic soil, but some of the things they owned or wore had made it down the centuries. One member of the lady's entourage had strange accessories formed from reworked Iron Age coins. Another was armed with a single-edged fighting knife. The lady herself had been adorned by gold jewellery, polished gemstone pendants and a matchless shield-shaped gold pendant with fifty-seven cloisonné cells surrounding a polished scallop-shaped garnet. The objects were made in the seventh century

(Plate 5). The woman was soon celebrated as a Saxon princess.[108]

The archaeologists were taken aback. They had been drawn to the site by an Iron Age enclosure near older ceremonial monuments. Early medieval funerals were quite often deliberately held beside prehistoric ritual features, but nothing like this ritualistic conformation of special dead had been seen before. However, while the Street House cemetery had no parallel, it did have a context. From around 600 a fashion arose for raising mounds and arranging cemeteries to commemorate dead leaders. Sutton Hoo (Suffolk) or Prittlewell (Essex) are famous for the range and richness of their goods, weapons and furniture. Others like Caenby (Lincolnshire) and Taplow (Buckinghamshire) were opened in the nineteenth century and poorly recorded, while yet more were pillaged without record in earlier times. Historians argue about what such barrows were for. Some see them as features of pagan-Christian transition, others as an expression of anti-Christian sentiment.[109] In either case there has been a tendency to link the arrival of rich barrows with a shift from sub-regional polities to full-scale kingdoms. Their European distribution and date, however, point to another explanation. Rich barrows are found around the margins of Francia, and they were being heaped up well after Frankish royalty and nobility had begun to bury each other in churches. Hence, while an imagined past suggested by older barrows may have been one motive for raising new ones, they also suggest 'a striving for a monumental expression of status, achieved in more developed cultures by means of funerary churches, above-ground sarcophagi, and tomb sculpture'. Their occupants have been described as 'competitive, insecure potentates, concerned to show themselves as good as their Frankish contemporaries and better than their English rivals'.[110] Such tombs have a special allure, partly because of their treasures but also because of the historical half-light in which we see them, part-way between legend and the age of written records. The treasures and heroes may have fed later stories of kings and knights asleep under hills.[111]

This noblewoman on a headland brings more seventh-century ladies and headlands to mind. Just down the coast at Whitby was Hild, that daughter of a Deiran prince who late in the 650s founded a monastery overlooking the mouth of the River Esk. Before settling at Whitby she had been a member of a religious community near the mouth of the River Wear, and then led a religious house across the Tees on the headland at Hartlepool. Her predecessor at Hartlepool was Heiu (fl. 640s), who also moved in magnate circles.[112] Around 655 Osuiu, king of Northumbria (c. 612–70), put his infant daughter Ælfflæd (654–714) into Hild's care. Ælfflæd succeeded Hild as abbess of Whitby, where for several years she ruled jointly with Osuiu's widow, Eanflæd (b. 626), daughter of the Deiran king Edwin. Looking further north, we are told of a religious community of aristocratic women close to the mouth of the River Tyne, and an abbess of Coldingham in the 660s who was half-sister of the Northumbrian kings Oswald (*c.* 604–42) and Osuiu.[113]

Royal abbesses were typically unmarried princesses, or widowed or separated queens. Many of them knew one another and moved between royal households in different kingdoms.[114] They exercised 'the authority and independence that were otherwise the prerogative of a king's wife'.[115] A number, including Hild's sister, had been members of religious houses in Francia and knew the kind of world which the barrow builders were trying to emulate. Their settlements on coastal frontiers and at river mouths became places where royalty were memorialised. With houses like Lindisfarne and Jarrow they formed a seaside monastic province.

A leading feature of this sphere was zeal for learning. Several houses of the princess-abbesses became centres of advanced teaching, writing and scholarship. Whitby under Hild and Ælfflæd trained five bishops, nurtured the first English poet whose name we know and produced the earliest biography of Gregory the Great to have come down to us – conceivably written by a woman.[116] Few objects evoke this newly literate world more strongly than small stone grave-markers engraved with names from Hild's earlier community at Hartlepool. They were found at different times during the nineteenth century, often in the course of mundane jobs like digging drains or laying foundations. In body they resemble a pack of A4 copy paper – quite small, chunky, the display surface flat and smoothly dressed, a little like a book cover. The inscriptions are piercingly artless: apart from a one-word name there is usually nothing more than a cross, sometimes coupled with an alpha and omega.[117] Read the women's names aloud – Hildithryth . . . Beorhtgyd . . . Torhtsuid . . . Hildigyth . . . Edlesuid – and they become whispers across the centuries.

It was in these exclusive circles of Deiran–Bernician nobility to which the princess overlooking Loftus and Carlin How had belonged. Deira and Bernicia are usually regarded as kingdoms that were merged during the later seventh century to form the greater realm of Northumbria.[118] Historians have accordingly considered them as territories and looked for their boundaries. Deira has been visualised as Yorkshire's forerunner, more or less co-extensive with the modern county, with a capital at York and a northern border along the River Tees.[119] Bernicia at full extent is equated with the region from the Tees up to the Forth. On this basis our princess was buried just inside Deira's northern frontier. However, a re-reading of Bede and other near-contemporary authors reveals that none of them used the terms Deira and Bernicia in a territorial way. Rather, they talked of the kingdom or region of the Deiri or the Bernicii. Bede also referred to the Deiri as a *tribus* – a category of people with a hereditary implication. In other words, he saw the northern kingdom as being divided between two peoples or dynastic groups.[120] (The *Hymbri* or *Humbrenses* who make up the second part of 'Northumbria' probably also referred to the people of Deira.[121]) On this understanding being Deiran or Bernician depended not on where you lived or died but on who your relatives were or whom you followed.

Yorkshire's origins, then, are probably not to be sought in some *Ur*-territory but in an elite extended family group. We do not know when the family was founded. A likely context would be a fifth-century power struggle between sub-regional groupings that came into being after the withdrawal of the Roman army, very possibly intensified by the prolonged worldwide volcanic winter that resulted from a violent eruption in AD 536. Some of these groups later saw themselves as British, others as Anglo-Saxon. Whatever they were, Bede's account of Northumbria's emergence seems to have covered only the last stages of the process, following a century or more of dynastic competition and successive amalgamation. Hence, while the fulfilment of Deiran family history was an English kingdom, its origin probably lay in self-identification. They took a name with ancient connotations: 'Deira' shares its root with words like *Derventio* (the Roman name for Malton) and Derwent, the British name of the river that flows out of prehistoric language and across a large part of Yorkshire.

The Loftus princess and Hild at Whitby were buried on Yorkshire's brink. Like Bede at Jarrow, Cuthbert on Lindisfarne, and the railway at Hunt Cliff, they were examined daily by the tides. For thirteen centuries the princess watched over Carlin How, and latterly oversaw Alf Myers and Uncle Fred as they tunnelled beneath her resting place. Carlin How, we recall, means 'hill of the witches'. Was part of the role of the princess and her retainers to deter the witches? Or could it be that some far-off garbled local memory of their graves gave rise to the idea that witches were in the vicinity?

2

TUNNEL VISIONS

Twenty years after the first visit to Skinningrove I was digging in York. Wintry showers rattled the heavy-gauge plastic canopy under which we worked. In December 1972 Access credit cards had been in use for just over a month, Chuck Berry's 'My Ding-a-Ling' topped the charts and Apollo 17 – the last manned trip to the moon – was preparing to launch.

Inner York is a knot of narrow streets bounded by two miles of medieval walls. Some of the streets have quirky names, like Jubbergate and Whip-Ma-Whop-Ma-Gate. Being in the history business we knew that 'gate' approximates to *gade* (say it 'gah-thuh'), the modern Danish word for 'street', and that this is because York is successor to the capital of an Anglo-Danish kingdom. Alleys and passages lead sideways off the streets. A newspaper proprietor called Edward Baines published a list of them in 1823.[1] The list ran to three pages and oh, what names! Mucky Peg Yard, Old Racket, Ogleforth – they lift the heart, and reading them gives a sense of what the city once was. Butchers' Lane, Tanner Row and Glovers' Passage tell of trades and crafts. Caroline Row and Ettey's Buildings recall people. Otley's Yard and Dougleby's Passage point to other places. Bedern, Petergate and Precentor's Court recall the clergy, choirmen and servants who lived around the cathedral. Jewbury remembers a medieval Jewish quarter. Gropecunt Lane (long since expurgated to Grape Lane) celebrates the workplace of whores. Gillygate, Clementhorpe and Sampson's Square, in contrast, contain names of saints. Giles, Clement and Sampson were among dozens of dedicatees of churches in city neighbourhoods to whom citizens looked for protection, with churchyards to receive them when protection ran out. The innermost parishes were close-packed; a good outfield cricketer could span one with a single throw and you can walk across the lot in half an hour. Such a walk is a zig-zag kind of journey because the streets are crooked. Turning corner after corner you pass through a succession of narrow, high-sided spaces from which the only long views are upward (to the sky, from Old Norse *sky*, 'cloud'). This means you can come upon a building like York Minster unawares – unless you happen to look along an alley and glimpse part of its flank, apparently floating, like some medieval airship.

York's walls and roofs are richly textured: rubicund bricks, white to pale yellow limestone, silvery oak, vermilion pantiles.[2] Architecturally the place is a multiverse; hardly any two buildings are alike. However, there are underlying rhythms. The typical unit of building is long and thin, with a house or shop on the street and a line of outbuildings of descending height projecting to the rear. Such parcels were laid out in the Middle Ages to a system that provided space for craft-working, brewing, storage, pigs and hens, tenements to let, or all of these and more at once. Later, some of them were subdivided while others were merged to form gardens or orchards which later still provided sites for small industries. By the early 1800s these included smoky iron-working concerns that turned out things like ornamental railings that went with Georgian and Regency houses.

By West Riding standards York in 1823 was not a very industrial or commercial place. Baines was accordingly sparing with the firms he singled out for special mention. One who made it was a maker of optical glass. A wholesale book company was another, apparently because Baines was impressed by it being 'among the first establishments of its kind' outside London. He continued: 'The city has also manufactures of carpets, linen, stuffs, flax, cordage, agricultural implements, combs, gloves, paper hangings, articles in chemistry, musical instruments and jewellery.' Glovers used to work on the street where we were digging. With them were leatherworkers who made things like scabbards, shoe soles, harness straps and dog collars. Scraps of leather turned up as we started work.

Alongside the gentrified York where bookworm citizens chose wallpaper, played violins and wore fine clothes ran the life of a town that served a farmed hinterland. Agricultural York was a place for trading beasts, butter and wool, where labourers came for hire or to gamble earnings after harvest, beer was brewed, carcasses broken, and farmers could buy tools or fix a cart. It revolved around markets and fairs. Such towns existed all over the country, but York stood out on account of its position.

The city sits centrally in the Vale of York where two rivers meet. One, the Ouse, is a main stem fed by tributaries flowing out of the Dales. The other, the Foss, rises in the Howardian Hills to the north. The rivers occupy ancient drainage routes. During the last glacial period they were overlain by an ice sheet. About 15,000 years ago the ice came to a halt just south of modern-day York, where it met the waters of a large melt-water lake trapped against North Sea ice. Debris that had been pushed by the glacier was deposited along the ice front in a low ridge, rather as a yard broom pushes debris into a line. As the ice wasted back more such ridges were formed, leaving a system of natural causeways across the Vale floor. The result was an area where land and water routes intersect. The Imperial Roman Army recognised this 13,000 years later when they selected it as the site for a legionary fortress. They called it *Eboracum* (more on names later), and when their new province of *Britannia Inferior* ('Lower Britain') was created around the end of the second century, Eboracum was the place from which it was governed. The Ouse was then navigable to the sea. In following centuries the successor place of *Eoforwic*, later Scandinavianised to *Jórvík*, flourished as an international port in a trading axis that ran from Dublin through York to the Baltic, along Russian rivers to Byzantium and Baghdad. Some of the dedicatees of York's churches give a hint of that range: the Norwegian king-saint Olave (995–1030), for instance, or Clement, patron saint of mariners on account of his martyrdom by being tied to an anchor and thrown into the Black Sea.

For us archaeologists these early contacts were witnessed by a mysterious mass of black, spongy soil that intervened between the level of the modern street and the fortress where Constantine the Great had been proclaimed Caesar. It was enigmatic because at first sight not much seemed to happen in it. Yet something had obviously happened for in places this stuff was eighteen feet thick. Its make-up gave the clue. Within it, if you rubbed it between your fingers or looked down a microscope, were things like seeds, flakes of skin, whiskers of dead cats, faeces, decayed timber, bits of buildings, ash, beetles, eggs of thread-worms, butchered bones and scales of fish. By 1972 archaeologists had come to realise that unread cultural histories were encoded in these dregs. ('Dregs', appropriately enough, comes from the Norse word for 'sediment'. Along with 'mire' (from *myrr*, 'bog'), 'rotten' (*rotinn*) and 'dirt' (*drit*, 'shit') they describe the tilth of organic debris in which York stands.)

Our presence on that particular site on that particular Thursday (Old Norse Þorsdagr, 'Thor's day') was another consequence of York's location. The Vale of York is one of only two main land corridors between

England and Scotland. The Vale provided a corridor of gentle gradients between York and the Tees valley, and thus a link between London, the north-east and Edinburgh. The first train from York to London ran in 1840. By 1860 a third of a million people were travelling the line annually. York was also a waypoint on routes to the West Riding and the east coast. Its position where main lines met suited the manufacture and repair of rolling stock, and the emergence of new industries, like production of cocoa and confectionery. Railways also brought tourists who revived fortunes that had dipped following the rise of the Ridings.

Motor vehicles reinforced York's expansion and heralded a new problem. From about 1900 the same geographical influences that had attracted medieval merchants and impressed Roman engineers made the city a place upon which cars and trucks converged in ever-increasing numbers. Following the upgrading of trunk routes between the wars, seven cross-country roads met in York's centre. By the mid-twentieth century central York was often at a standstill.

Everyone agreed that York's historic core was important and that the exclusion of cross-city traffic was necessary to protect it. There was less accord on how to do it. In the 1950s and '60s the standard solution was to build a ring road to conduct through traffic around the centre while enabling local traffic to reach all parts of the town from a series of roundabouts.[3] This had been done in other cathedral cities such as Chester, Canterbury and Hereford.[4] The result often turned out to be a noose rather than a relief, but by the late 1960s such costs were not yet clear. At Worcester a new relief road had even encroached on the cathedral close – a reflection of the then-prevailing belief that accommodating vehicles came before anything else. From that it was not far to an assumed binary opposition between progress and retreat in which caring for existing surroundings was associated with retreat. Thus it was that on 3 August 1971 York City Council voted by roughly two to one in favour of building a dual carriageway around the historic centre.

Others said that York's main asset was York and that the city's intactness as a special place should take precedence over the rights of haulage companies to route HGVs through its medieval streets. Those who said this added that the ring road would be a noose. building three miles of four-lane highway around the city walls would do away with good buildings, mar approaches to medieval gates, and carve through areas where green spaces, Georgian and early Victorian inner suburbs were happily mingled. Such a road would split York into a cossetted core, inharmoniously girdled by concrete, roundabouts and flyovers, and an outer realm of housing and industry that by implication was of no interest.

Nearly 9,000 residents formed a group to oppose the road. Looking to the future, they called themselves York 2000. Meanwhile, a new charity, the York Archaeological Trust, was created to salvage evidence of the city's past ahead of the expected upheavals. During the summer of 1972 the first of their staff appeared and American student volunteers arrived to help them. Meanwhile, land earmarked for the road was blighted and buildings along its corridor were run down or compulsorily purchased. Opinion polarised, tempers frayed. York 2000's leaders, according to a city councillor, were mainly 'from or on the fringe of the university who are usually described, loosely but never in my opinion accurately, as intellectuals'.[5] Whoever they were, they sold keyrings, mugs and biros, and published a cookbook, to raise funds to hire independent consultants to evaluate the scheme and give evidence at the public inquiry which had been called to assess whether or not the scheme should go ahead.

The inquiry opened in mid-October 1972.[6] The arguments made during its six weeks have since become the subject of books.[7] Looking back, the controversy about York's inner ring road can be seen as a watershed in how we value our surroundings. It did not feel so momentous at the time. We were well aware of the dispute, of course – in the presence of so many shabby, boarded-up buildings and local newspaper headlines it was impossible not to be – but we were preoccupied with our own work under York Minster, excavating in support of a programme of emergency repair. Moreover, as autumn turned to winter the student diggers went back to their studies and the big excavations had wound down. By the time the barristers at the inquiry made their closing submissions things seemed to have settled. Then, in late November, a message arrived from the York Archaeological Trust. They were not yet at full strength and events on a site elsewhere in the city led them to ask us for help.

Steve, a colleague, and I went to look. It was a corner site, between Church Street and Swinegate, and contractors were preparing it for new shops and offices. The area lay inside the former fortress so the presence of Roman structures ten to fifteen feet below the street was more or less guaranteed. Back then, however, there was no consensus that the public interest would be served by enabling archaeologists to examine such sites before developers built on them. Indeed, in this instance a request for access between demolition of the old buildings and the start of development had already been refused. Poetic justice thus attended what followed: as the contractors began to drive piles through the floor of an old cellar they met an obstacle. A mechanical excavator was brought in to expose the problem. About eighteen feet below street level the machine hit something large and hard. The hindrance turned out to be a gritstone slab about the size of refrigerator and weighing about three tonnes. The digger driver prised it out with the back actor of his machine. Beneath it was a void.

Steve and I peered into the pit. The sides had been roughly shuttered with boards. At the bottom was a black rectangle where the slab had been torn out. A few yards away a Roman bath house was emerging from another mechanical excavation. It was with this that the Trust had asked for our help. While that was being organised Steve and I talked to the Trust archaeologist who had been assigned to the discovery.

As a child I had read stories about secret passages. Here was a real one. Steve descended the ladder. The tunnel at this point was not very large: about one and a half feet wide and three feet deep. Steve stepped into the channel, put his elbows on the side walls, and let himself down. When prone, he reached for the heavy-duty torch and crawled away into York's netherworld.

He was gone for maybe fifteen minutes. When he reappeared he was covered with greyish-brown dirt and grinning. 'It's a Roman culvert system.' He gave me the torch, and some advice. The first few yards would be tricky: the channel narrowed, there was a chicane and silting had reduced headroom to a few inches. But I should press on – it would get easier. There was an alignment on the left that I should ignore, and I should likewise disregard a further channel that would be facing me on the way out.

I climbed in, lay down and shone the torch. What little I could see did not look promising: a ramp of silt and mud rose towards the roof. As I wriggled forward there was less and less space; at the chicane the channel was almost completely silted and the further I went the tighter it became. After a few yards I was barely able to advance, powerless to reverse, and incapable of turning. Was I being entombed? I tried to concentrate on Steve's advice: keep going and it will be OK. Eventually I rounded a corner. The clogged channel opened into a taller, wider, well-built passage, walled with long rectangular stones and roofed with massy blocks. After the panicky crawl this space felt safer; it was even high enough for a stooping walk. I stood up and set off. Not far along was an arch formed from long wedge-shaped stones. I ducked to pass through. Torchlight showed a similar arch about thirty yards beyond.

10. Roman culvert under excavation, York, 1972

Just ahead was a side passage. It was about two yards long. I wriggled in and looked up: above was a circular opening in the roof. Back in the main alignment, I walked on. More side passages led off to left and right in turn. Several of them were vaulted. The roof of one of them had collapsed. I reached the next arch and ran a hand along its surface. Presumably the vaults had been positioned to bear the weight of walls that had run overhead. I was beginning to get the idea: not only was I walking about in subterranean spaces which no one had entered for 1,600 years, but their features provided a diagram of the Roman fortress above.

Work on the development was stopped for a few days while the Trust archaeologists surveyed the system and did what they could to examine its contents. They ran power cables into the passages, rigged lights, recorded the structure and took samples of silts and fillings for sieving (Fig. 10). The system was elaborate. There was a main channel, accessible for 144 feet, with which six side passages connected. The chief alignment linked with more channels, although modifications and diversions in Roman days, coupled with more recent disturbances, precluded exploration to its full extent. The signs were that the system had served a large bath house and discharged into the tidal River Foss.[8]

Within the culvert's silts and fillings were lost objects.[9] Gaming counters and bone needles for the repair of clothes were typical of soldiery on stand-down. Intagli of red jasper, cornelian and chalcedony bore images of luck-conferring divinities like Mars, Roma and Fortuna. Gold pendants, an earring and hairpins hinted at women in the fortress. But what were we really to make of such items? For most of its several-hundred-year life the system had been kept clean, which meant that the majority of the things in it had accumulated after the culvert went out of use and cleaning stopped. Here and there were smidgeons of later material, as at one place where bits of clay pipe and the bones of a dog marked an occasion back in the 1840s when men digging a well inadvertently broke through the roof.

The culvert's interior offered a kind of reverse perspective on other places.[10] The stones used to build it came from at least four distant parts of Yorkshire. Bones of migratory fish like eel and smelt pointed to a time when York's rivers had been tidal. Microscopy revealed strong pollen signals from oak, lime, elm, grasses and heathers. This was at odds with evidence from sites on the surface which said there had not been much woodland around Roman York. The tree pollen must have come from elsewhere, perhaps in water ducted from springs in wooded districts that were miles away – at the foot of the Wolds, maybe, or the Howardian Hills, with moorland nearby. For distance, however, nothing approached a tiny piece of silk that was found while sieving for bits of insects and plants. The weave was typical of cloths made up in Syria; the thread had been imported from China.

The culvert, then, was a kind of lens that brought faraway things into present view. More than this, it did so in ways that contradicted normal rules. Archaeologists work on the basis that units of stratification are governed by laws – for example, that upper layers must be more recent than lower ones, or that cultural objects cannot be younger than the strata in which they occur.[11] The culvert seemed to be outside such laws. The ceramics specialist noticed that the find-spots of pot fragments which fitted together were spread far apart and that some of the sherds occurred in layers where the rules said they had no business to be. The reason was that objects introduced through inlets and drains had been bowled along for long distances, and that when they finally came to rest the sediments had sometimes been so runny that they sank through pre-existing layers.

The theme of the subterrane as an alien environment introduces other things in other types of place. Yorkshire is aerated by all kinds of chambers and regions beneath the ground. Seekers for coal and iron, lead, copper, alum, potash, gypsum, whinstone and jet left tunnels that run for scores of miles. With them are stranger spaces, like the medieval hermitage hacked downwards through rock beneath Pontefract at the end of the fourteenth century, or the bunker constructed on the out-skirts of York the year before the Cuban missile crisis.[12] In the event of nuclear attack a small team of specialists would have congregated here to monitor fallout, living on for a few days in a burned-out, poisoned world until their means to filter air and generate power ran out.

Apocalypse of another kind is suggested by tales of knights at rest beneath hills awaiting their country's hour of need. Arthur and his war-riors are said to sleep in a cavern within Round Howe at Richmond. According to the story a potter called Thompson was led into the cave where a horn and sword were offered to release the sleepers from their enchantment. As Thompson began to draw the sword from its scabbard the sleepers stirred. Fearful, Thompson returned the sword and fled.[13] This belongs to a group of such stories. At Sewingshields near Hadrian's Wall a local farmer found his way into a hall wherein slumbered Guine-vere, Arthur and his court. Before them lay Arthur's hunting hounds, dozing on the floor beside a fire which burned without fuel.[14] If roused at the wrong moment the charm cannot be lifted until a given time has passed. Freebrough Hill, on the edge of the moors not far from Skin-ningrove, is an 800-foot-tall conical mound. It has been described as the 'Silbury of the North' but is in fact a geological feature. Even so, legends and romantic ideas swirl around it. Among them is the suggestion that Arthur and his knights lie inside. Another says that the hill takes its name from a dedication to Freyja, the Norse goddess of love, sex and battle. In 1661 a traveller on his way from Whitby to Guisborough was told by locals that it had been 'cast up by the Devil, at the Entreaty of an old Witch, who desired it, that from thence she might espy her Cow in the Moor'.[15]

People-under-the-hill stories are found all over Europe, notably in parts of Germany, Austria and Scandinavia.[16] They have deep roots. The story of a cup made of strange metal seized by a passer-by from men and women who were feasting inside the mound of Willy Howe on the Yorkshire Wolds was written down in the twelfth century.[17] The legend of 140 knights beneath Alderley Edge (Cheshire) waiting to fight the last battle of the world has been taken back to the Bronze Age.[18] Arthur at Richmond recalls the belief that Emperor Frederick Barbarossa (1122–90) is inside the Kyffhäusergebirge, a range of hills between

Thuringia and Saxony-Anhalt.[19] Barbarossa died while attempting to cross the Saleph river in Armenia during the Third Crusade. Since the fourteenth century, legend has insisted that he lives on in the Kyff-häuser, where his continuing presence is signalled by ravens that circle the summit. This possibility was no doubt helped by the remoteness of Barbarossa's real remains, which were divided between several sites up to 1,500 miles away.[20] Arthur was similarly off the scene, his very existence uncertain and his grave at Glastonbury the result of a twelfth-century swindle.[21] Like the Virgin Mary such incorporeal figures lend themselves to local adoption in more than one place.[22] They are a bridge between a golden age and the end of time. Yet they are nowhere near as powerful or strange as things in caves beneath the mountains of Craven.

Craven is a district that spans the Pennine watershed, from the edge of Bradford almost to Lancaster and the Lake District, encompassing the sources of the rivers Ribble, Wenning, Aire and Wharfe. Craven's higher hills and mountains stand to the north. They are made of lime-stone, which we have seen was formed 350 million years ago when Yorkshire was on the bed of a sea close to the equator. The caves are more recent. Their formation began around 2 million years ago, when large outcrops of limestone were exposed and rainwater began to per-colate into natural joints and clefts in the bedrock. Since rainwater is mildly acidic the limestone gradually dissolved along its seams, creating a network of subterranean voids. These spaces were further developed by the effects of alternating glacial and warm episodes. During warm periods and spells of deglaciation when meltwaters were released the eroded fissures were enlarged and extended by the scouring of water-borne sands and stones. When the ice came back valleys were made deeper, enabling subterranean streams to seek lower channels, leaving a network of dry remnant caves above.

Craven's cave systems and underground streams and rivers have their own lexis of pots, gills, holes, swallets, passages and sinks. There is fancy in the naming of entrances like Nippikin Pot, Rumbling Hole and Wretched Rabbit. Some of the spaces to which they lead are vast. Gaping Gill is far deeper than York Minster is tall.[23] Rowton Pot is greater. When a cave becomes unduly large its roof may fall in, so turn-ing the space into a crag-sided gorge. Many caves connect. The Three Counties system runs for at least fifty-five miles, enabling the commit-ted caver to go underground in Cumbria, travel beneath Lancashire and return to the surface in Yorkshire. Michael Drayton recorded local belief about this netherworld:

As up towards Craven hills, I many have of those,
Amongst the crany'd cleves, that through the cavern creep,
And dimbles hid from day, into the earth so deep,
That oftentimes their fight the senses doth appal
Which for their horrid course the people Helbecks call.[24]

Caves defamiliarise things we think we know. Life, colour, sensation and sound are made strange. Insects, flatworms, fish, bats and sightless crustaceans live in their different neighbourhoods. Rock surfaces normally jagged or flat are puckered and rippled. Exotic structures and surfaces have been created by dripping water. A droplet that falls vertically deposits a minuscule circlet of calcite. Over scores of years such rings accumulate to form tall water-filled tubes. The tubes are akin to straws, fragile, likely to snap at the faintest contact, but if they block or water flows down their sides they may fatten to form larger columns. Minerals dissolved in seeping water produce flowstone, which can take shape as translucent drapery, a frozen waterfall or some fantastic chandelier of tubal clusters.[25] If you tap a flowstone pillar it is likely to give out a pure-toned ring. Sound is heard differently to the open: in deep caves the only movement of air is caused by the visitor.

Dry relict caves became collecting places for archaeological and natural deposits. Such residues are evidentially special because they were sheltered from the effects of later glaciation. Victorian antiquaries and geologists realised that animal bones found inside Craven's caves could throw light on changing climate. Victoria Cave, in Ribblesdale near Settle, became – and remains – a centre for such inquiry. The cave's main chamber was discovered in 1837 and received its name following the accession of the new queen in June that year. The formation of the Settle Cave Exploration Committee led to sustained study (Fig. 11). Excavations in the 1870s disentangled a profuse deposit of bones from animals that had been hunted or scavenged by hyenas. Among them were now-extinct species of elephant and rhinoceros, hippopotamus, bison and giant deer. Wolves, reindeer and brown bears that died during hibernation suggested colder times.[26]

Modern methods enable us to assign dates to these remains with a preciseness that was not available to Victorian cave scientists. We now know that deposits inside Victoria Cave were formed across more than 600,000 years, during which there were four great climatic swings. The heyday of the great beasts – the hippos, the elephants, the rhinos – was a temperate interlude roughly between 130,000 and 25,000 years ago,

11. Victoria Cave, entrance to the Settle Cave Exploration Committee's excavations, 1870. The photograph was one in a series – apparently the first systematic use of photography to record an archaeological cave excavation in Britain.

when much of Craven was open grassland interspersed with woodland. Then the ice came back and Craven reverted to polar desert. Warming resumed *c.* 15,000 BC. Within 500 years herds of reindeer and horses were moving into the area, followed by people who hunted them.[27]

For the rest of prehistory – about 14,000 years – Craven's caves seem to have been used at some times and avoided at others. Hunter-gatherers explored them in the sixth millennium BC. The presence of periwinkle shell beads in several caves suggests that they ranged afar, for the nearest source of periwinkles is the shore of Morecambe Bay, over thirty miles away.[28] During the Neolithic period some caves seem to have acquired specialised functions. One was used for sequestering human skulls, another for skulls and smashed cattle bones; at North End Pot a mace head made of red deer antler was deliberately left perched on a ledge about a hundred feet down an entrance shaft.[29] Here and there shafts acted as pitfall traps in which wild beasts perished;[30] crevices were selected to receive human burials.[31]

The heyday of leaving things beneath Yorkshire began towards the end of the first century AD, following the arrival of the Roman army. For the next three centuries an extraordinary range of goods was deposited underground in Craven – jewellery, weapons, tools, bits of vehicles, harness gear, cosmetic instruments, coins, domestic fittings. Since it is mainly metal that survives it may well be that there were also organic things like clothes or food that we no longer see.[32] The objects suggest a preference for caves on the fault scarps near Settle and for the Victoria and Attermire caves in particular.[33] Some of them, like small decorated spoon-like objects with spiral shafts and perforated bowls, or circular copper alloy brooches bearing triple spiral patterns, occur in such numbers that it is difficult not to suppose that they were made to be sequestered. One fragment from a pot that was made early in the second century bears a scratched name – Annamus. The name comes from the Roman province of Noricum, which corresponds with parts of present-day Slovenia and Austria, and perhaps more relevantly to a Roman imperial mining region noted for its output of silver, iron and lead.[34] There are few direct signs of Roman lead mining in the Yorkshire Pennines,[35] but stamped ingots lost in transit confirm that such production took place in a highly organised way. The fact that only a few of them have entered the modern record simply reflects the alacrity with which they would have been recycled by earlier finders. In any case, lead was used in prodigious quantities in Roman Yorkshire, for pipes, tanks, roofing, construction, salt pans and coffins.[36] Bringing the different kinds of evidence together – and remembering Drayton's reference to underground streams as 'hell becks' – it is not difficult to arrive at a reading in which prospectors and members of the military from across the Empire held the caves to be places in which to make offerings to ancestral spirits and the *di inferi* ('the gods below'). The notion of a link between the recurrent triple-spiral pattern and the 'three ways' of Hecate, Luna and Proserpina might take this too far, although if we were to look for a goddess of climate change, Proserpina's abduction into the underworld and subsequent association with the cycle of winter and spring make her an attractive candidate.

Yorkshire has been scrabbled, tunnelled, prodded and half emptied in search of different minerals for up to 4,000 years. Nearly all the workings are now forgotten, yet it is pits, adits, drifts and their spoil that give much of the county's face its distinctive expression. Few visitors to the Yorkshire Dales National Park realise the extent to which its landscape has been shaped by producing three-quarters of a million tons of lead between 1845 and the Second World War. The pits and rakes are now clad in heather and bracken; hushes and waste heaps are increasingly mistaken for natural features.[37] Sharp-eyed motorists passing Wakefield on the M1 may notice blister-like bumps in fields to either side. They are the traces of early shafts, hand-dug in search of coal, each extended sideways until the space approached collapse, then abandoned and another begun.

Jet is a variant of coal. The Tudor antiquary William Camden called it 'black amber'. On Yorkshire's coast, he wrote: 'It groweth among the cliffes and rockes where they chinke and gape asunder. Before it be polished, it is of a reddish colour, but after it be once polished it becometh, as saith Solinus, as a Gemme of a bright radiant blacke colour.'[38] Nine centuries earlier, writing in Jarrow, the Anglo-Saxon scholar Bede wrote that jet will burn if put into a fire and drive away serpents when kindled. Bede noted that if jet is warmed by friction 'it attracts whatever is applied to it, just as amber does'.[39] Jet is the fossilised wood of a close relative of today's monkey-puzzle tree that grew around 180 million years ago and came to rest in the anaerobic sediments of a Jurassic sea. The trunks have been flattened under compression to individual spars with the section of a lens. Their volume is tiny in relation to the strata in which they occur, so jet mining has always been hit-and-miss. The earliest jet jewellery in Yorkshire is found with Bronze Age burials. Jet has been prized ever since, but the heyday came in the second half of the nineteenth century when fashion, customs of mourning (boosted by high infant mortality during the Industrial Revolution) and royal patronage combined to turn a Yorkshire craft that employed a handful of people in 1832 into a major industry centred on Whitby by the 1870s. It used to be thought that the mines dug into the flanks of the North York Moors were mostly simple drifts, no more than the width and height of a man, close to the surface and seldom much longer than 300 yards. The teams that worked such mines were accordingly visualised as single families, small groups of part-timers or redundant ironstone miners. Timber supports were few to keep costs down; lighting might be a candle in a tin or stuck to the wall with handful of clay. Around 120 such openings are known today; landslides occasionally uncover more.[40] Exploration of the workings themselves shows that the size of the industry has been underestimated; while small workings were indeed common, mines with passage systems extending over three miles (five kilometres) have now been surveyed.[41]

The idea of the buried space as a place of outward vision brings us finally to Boulby, between Loftus and Whitby, finally not least because it is the last working mine in Yorkshire. Boulby's main business is the extraction of potassium chloride – potash – from the crystallised salts of the Zechstein Sea that finally disappeared 225 million years ago.[42]

Thirty-seven square miles of deposit have been withdrawn since the mine was sunk in 1968–9. This is a modest figure when set against Boulby's contribution to how we understand the universe.

For eighty years or so it has been clear that there is more to the universe than we can see. Suspicions arose in the 1930s when astrophysicists began to calculate the orbital velocities of stars in the Milky Way and other galaxies. Their calculations, based upon the gravitational effects of such things as the observable mass of stars, dust clouds and gas, suggested that a very large amount of mass was missing. 'Dark Matter' was proposed to account for this absence – 'dark' because it neither emitted nor absorbed electromagnetic radiation (light, X-rays, gamma rays). The first direct evidence for dark matter was obtained from galactic rotation curves in the 1970s. More indications were obtained from gravitational lensing (a prediction of general relativity that gravity can bend light rays) and from other cosmological observations.

In the 1990s it was confirmed that the universe is expanding. However, contrary to the expectation that the rate of expansion should slow as time passes, it was found that the rate of expansion was actually increasing. The only plausible explanation was that in parallel to dark matter the universe also contains 'dark energy'. Dark energy appears to constitute 68 per cent of the known universe. Calculations indicate that dark matter provides a further 28 per cent, meaning that the universe we can see and understand is only 5 per cent of what is really there.

We do not know what dark energy and dark matter are or what form they might take. The nature of dark matter, however, is a little easier to postulate. It is believed to be some form of exotic fundamental particle. Most known fundamental particles (like pions and muons) produce some form of radiation when they interact with real (baryonic) matter. In the case of dark matter the fact that we do not observe this radiation implies that it is made up of very weakly interacting particles. Computer simulations suggest that these weakly interacting massive particles may have more than 200 times the mass of a proton – almost the mass of a uranium atom. The experiments at Boulby are, in part, to detect them. Boulby was chosen for the purpose because the detectors must be shielded from other forms of background radiation, such as cosmic rays and natural radioactivity from rocks. At 0.68 miles (1.1 km) Boulby is the deepest mine in the UK, which reduces the cosmic ray background to a millionth of what it is at the surface. Boulby is also a salt and potash mine, with deposits that we have seen resulted from the evaporation of a primordial sea. The experiment is housed in a rock salt cavern, and rock salt is low in natural radioactivity. Thus it is that a place beneath Yorkshire (and coincidentally next to Carlin How) has become a point of universal outlook. The work of the Settle Committee back in the 1870s was akin to it: revolutionising inquiry, in that case into deep time, that introduced new ways of understanding the world, and our place within it. Like the nearby (and contemporary) Ribblehead railway viaduct, Victoria Cave witnesses the exhilarating progress of mid-Victorian science and technology. Tunnels, then, are windows into the future, past, and imagination.

Alongside King Arthur at Richmond is another fable with an underground theme, about a passage said to run nearly a mile from Richmond Castle to Easby Abbey. In the days of George III some soldiers attempted to trace the path of this tunnel by putting a drummer-boy into it and following the sound of his drum. This worked for a distance; then the beat stopped. The boy was never seen again, although the tale records that on some evenings residents of the town hear the tap of a drum.

This legend came to mind as I was crawling out of the Roman culvert, flat on my belly in muddy silt. Could it be that earlier chance discoveries of stretches of York's Roman culverts had given rise to the similar story of the Blind Fiddler? Long before York was besieged by cars and trucks it was invested by the Parliamentary and Scottish armies during the Civil War. Rumour had it that people were being smuggled in and out of the city through underground passages. They were escorted by a blind man who knew the system. The tunnels were unlit, but the blind man played a violin, and the people he led followed its sound.

The siege of York was lifted in July 1644 following the Battle of Marston Moor. The later siege by inner ring road was ended in February 1975 when the report made by the inquiry inspector (who had found in its favour) was overturned by the Environment Secretary, Anthony Crosland. Here was another thing: Marston Moor, Adwalton, Towton, Boroughbridge, Stamford Bridge, Fulford, Hatfield – traffic in towns: why was it that for well over a thousand years so many key battles were fought in Yorkshire?

3

DERE STREET

A round 1131 Henry, archdeacon of Huntingdon (*c.* 1088–*c.* 1157), completed a history of the English. In it he wrote about four long-distance highways. These roads, he said, were old. They had been made under royal authority to deter enemies by allowing swift movement of armies. Each had a name. One ran from east to west and was called 'Ichenild'. Another was called 'Erninge' (sometimes 'Ermeninge') and went north–south. 'Watling' connected Dover with Chester and 'Foss-way' ran from south-west England up to Lincoln.[1] A tract about the king's peace written around the same time also spoke of four 'ways': Watling Street, Fosse, Icknield Street and Ermine Street. Two ran longitudinally, the other two crosswise.[2]

The cleric and fantasy historian Geoffrey of Monmouth (d. 1154/5) ascribed the four roads to a fabled king of the British back in what we would now call the Iron Age.[3] In fact, they were built by the Romans, and wherever the names came from they are first met in writing in the ninth and tenth centuries.[4] Moreover, confusions arise because some of the names were used for more than one road. There is an Icknield *Way*, for instance, which runs from Norfolk to the Chilterns,[5] and Icknield *Street* that runs northwards from the south Midlands. In 1344 Ranulf Higden (*c.* 1280–1364), a monk writing in Chester, repeated Geoffrey of Monmouth's claim that this royal road came from west Wales to Worcester and thence to the mouth of the River Tyne by way of York. Variant names occur in different manuscripts, but the basic form used by Higden was Ryknild.[6]

Ryknild Street branched north from the Fosse Way in Gloucestershire. It connected Roman towns, staging posts and forts preceding today's Alcester (Warwicks.), Wall (Staffs.), Little Chester and Chesterfield (Derbys.) on its way to *Danum*, Doncaster, where it was joined by the western branch of Ermine Street that came up from Lincoln. From here the name fades, but the road continued to Pontefract and

Castleford, following the limestone ridge on the western edge of the Vale of York to avoid the Humberhead marshes. At Bramham Moor the road turned north-east for York, whereafter it returned to its northerly track via the town of *Isurium Brigantum* (today's Aldborough, 'the old burg', forerunner of Boroughbridge). The dog-leg made two sides of a triangle, so wayfarers who wished to go north or south without passing through York were provided with a short cut along the third side (Fig. 12). This stretch is called The Rudgate, and when I was digging in York in 1971 I lived two miles from it.

Newly married, we rented an agricultural labourer's cottage on the windy corner of a farm called Ingmanthorpe Grange. Ingmanthorpe is just north of the former coaching and posting town of Wetherby and a mile east of the A1 trunk road. The A1 was the twentieth-century successor to the Great North Road, and the Great North Road between Wetherby and Doncaster was in its turn the heir to Ryknild Street. If Michael Drayton was writing now very likely he would anthropomorphise the A1 as some sort of spirit or dryad. The spirit would have a rough story to tell.

But to set the scene: Ingmanthorpe is farmland; there is no village, just a hall and a couple of farms. It may always have been like that. The name combines the Scandinavian given name *Ingimundr* with *thorp*, the Old Norse word for a small settlement or farm. If you take the given name to pieces the 'Ing' means something like 'the name of a god', and 'mund' translates as 'protection'. Who this particular Ingimundr was we'll never know, but the place was carrying his name by the late eleventh century, when it was noted in the Domesday survey.[7] A century later the manor was in the hands of a cadet branch of the baronial de Roos family, who retained it until the barony passed to the earls of Rutland in 1512.[8] Robert Roos (1510–83) held the manor from them, and had other property. However, he also had debts and became embroiled in litigation. One by one he sold or surrendered his estates and passed his last years in obscurity. Ingmanthorpe passed to William Cecil (1520–98), first Lord Burghley, Lord Treasurer to Elizabeth 1. Cecil and his successors let it to tenants.[9]

In 1971 Ingmanthorpe Grange was connected to the wider world by a single-track road called Loshpot Lane. Grass grew along the lane's centre and so few vehicles used it that the farm dog could recognise the engine noise of its owner's car nearly a mile away. Going east the lane came to a dead end; heading west it connected with the southbound carriageway of the A1 beside the forecourt of the Midway filling station.

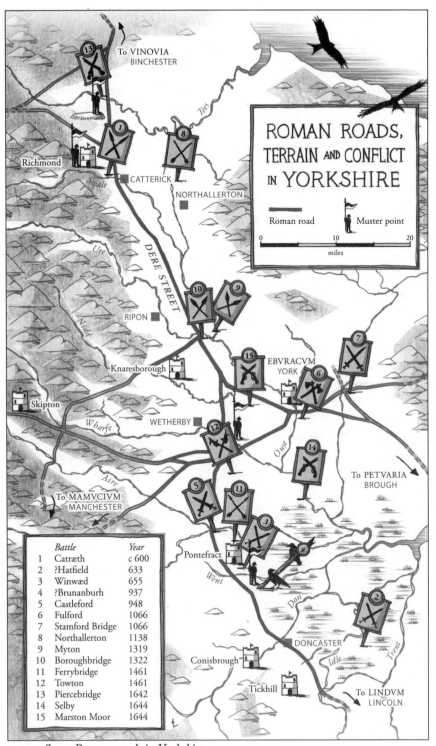

To VINOVIA
BINCHESTER

Tees

Richmond

CATTERICK

NORTHALLERTON

Swale

Ure

DERE STREET

RIPON

Nidd

Knaresborough

Skipton

Wharfe

WETHERBY

EBVRACVM
YORK

Ouse

To PETVARIA
BROUGH

Aire

To MAMVCIVM
MANCHESTER

Pontefract

Went

Don

Conisbrough

Idle

DONCASTER

Trent

Tickhill

To LINDVM
LINCOLN

ROMAN ROADS, TERRAIN AND CONFLICT IN YORKSHIRE

Roman road   Muster point

0   10   20
miles

	Battle	Year
1	Catræth	c 600
2	?Hatfield	633
3	Winwæd	655
4	?Brunanburh	937
5	Castleford	948
6	Fulford	1066
7	Stamford Bridge	1066
8	Northallerton	1138
9	Myton	1319
10	Boroughbridge	1322
11	Ferrybridge	1461
12	Towton	1461
13	Piercebridge	1642
14	Selby	1644
15	Marston Moor	1644

12. *Some Roman roads in Yorkshire*

The Midway was a rectangular glass-sided box that stocked sweets, newspapers, and a few motor spares like fan belts and bulbs. It took its name from an inn called the Old Fox that stood a couple of hundred yards further up the road. The Old Fox had previously been the Halfway House, because it stood midway between London and Edinburgh. There was once a milepost at the foot of a nearby tree in which hung the inn's sign. Just to the south was the place where the Great North Road had emerged from Wetherby before the bypass was opened in 1959. It was called Deighton Bar, an echo from turnpike days when travel along the road had been controlled by toll gates and side bars.

The hum and rumble of A1 traffic could often be heard in Ingmanthorpe. Its noise varied from day to day. We soon learned that the sound was governed by the weather. If it was distinct then rain was in the offing; silence meant a wind blowing from the north or east, which in winter presaged cold or even snow. It was as if the road had moods. Its grumpiest days were around Bank Holidays or race days, when traffic backed up from the Wetherby roundabout. The Friday of Whitsun weekend was busiest; this was the day when frustrated drivers hoping to outflank the jam appeared in Loshpot Lane and so created a jam of their own.

For sheer unlikelihood, however, nothing beat the routine of Mr Dalby's dairy herd. This began around nine o'clock each morning when farmhands stepped nimbly into the traffic on the A1 just south of the Midway, flagged drivers to a halt, and ushered a procession of Friesians across the carriageways. After milking they did it again in reverse. The animals knew to move fast, and after several years the Department of Transport installed warning signs consisting of the silhouette of a cow and some alternately flashing lights. Despite this precaution the twice-daily parting of the A1's flow seemed at least as improbable as the Israelites' crossing of the Red Sea.

The 'Losh' in Loshpot Lane came from eighteenth-century slang for liquor, presumably because it took you to the Halfway House. The inn's successor, the Old Fox, was fondly regarded by locals and travellers, but dualling of the A1 had made it awkward to reach on foot, while the introduction of drink drive legislation in 1967 discouraged patrons arriving by car. The landlord and brewery bravely embarked on another makeover in which the Old Fox became a Tyrolean gasthof with overhanging eves, balustraded balcony and a shiny interior of varnished pine. They called it the Alpine Inn, and the licensee who presided over these incongruities was an impresario who installed a powerful electronic organ and hired entertainers like Ronnie Hylton and Roy Hudd. For a time, the Alpine roadshows were a draw, and on nights when there was a crowd in and the wind was in the west the throb of the organ could be heard at Ingmanthorpe. Then the A1 was again upgraded, this time to a six-lane motorway, and placed on a new alignment that left the Alpine forsaken. It closed, and in 2005, the year that the new motorway opened, The Alpine Inn burned down.

Similar tales of heyday and demise can be found along the path of the superseded Great North Road, where sights like The George at Catterick or the filling station in Leeming Bar once stirred expectancy among travellers who were heading north or a sense of homecoming on return. Georgian inns and larger farmhouses north of Wetherby had a distinctive look: brick-built, broad, sometimes double-piled, chimneys at the gable ends, and generously lit by tall eight- or twelve-pane sash windows. Many of them have gone; others are haggard or boarded up beside redundant bits of carriageway. Gone, too, from our journeys are mellifluous and quirky local names like Quernhow, Ainderby Steeple and Melmerby: the A1M no longer has local exits and hurries travellers through an anonymised countryside.

The Great North Road between Boroughbridge and Scotch Corner followed the Roman road from the legionary fortress at York to Hadrian's Wall and the mouth of the Tyne. In the Middle Ages this section was known as Dere Street (the 'Dere' from Deira), later as High Street or Leeming Lane. Mail coaches used it from the later eighteenth century, when it was levelled for coach traffic linking cities like Leeds and Newcastle. By the 1830s 3,000 coaches clattered along Leeming Lane each year. Among them were regular long-distance coaches with names like Express, North Star and Defence. Boroughbridge was a place of connection between river and road; for a time it became one of the busiest places on the route, serving travellers, boats carrying cargoes of lead from the Dales, or luxuries to them, and drovers who daily moved hundreds of cattle to southern markets. Then came the railway, and the coaches, boats and drovers went out of business.[10]

At Scotch Corner there is a parting of the ways: the A66 branches north-west for Penrith and Carlisle, while a mile further on the A1 curves eastward for Darlington, leaving Dere Street to carry on as a quiet secondary road through open country.[11] Until the 1930s Scotch Corner was a crossroads flanked by an eighteenth-century inn called The Three Tuns. In the late 1930s the crossroads was replaced by a roundabout, and the inn was pulled down to make way for a dour neo-Georgian

13. Scotch Corner: hotel and newly laid-out junction between A1 and A66 in 1938

three-storey brick hotel which was cleverly set at an angle to the junction and given canted corners, so that it always appears to be facing you if you approach it from different directions (Fig. 13).

Dere Street, in contrast, becomes peaceful (Fig. 14). A southbound traveller here is granted one of the finest of all first views of Yorkshire: a panorama stretching sixty miles from Cleveland to the western fells. Yorkshire's actual threshold is at Piercebridge, where the village green on the Durham side of the River Tees overlies the Roman fort of *Morbium* that watched over the crossing. The present bridge is a few hundred yards upstream from its Roman predecessor and the road jinks sideways to align with it.

On Monday, 1 December 1642, the bridge was the scene of a skirmish. The encounter was precipitated by events thirty-eight days before in Warwickshire, when the inconclusive outcome of the Battle of Edgehill forced Charles I and Parliament to plan for a longer war. During the summer the King appointed William Cavendish, Marquess of Newcastle (1592–1676), as his general in the north. Newcastle was described by a contemporary as having 'a tincture of romantic spirit' and the outlook of a poet. He had little military background, but selected able commanders and applied his wealth to the maintenance of his forces. Among them were three regiments recently raised in Durham and Northumberland. They were commanded by John Hilton, Sir William Lambton and Col.

14. Dere Street

Posthumous Kirton, and uniformed in white coats with crosses of blue and red silk sewn onto the sleeve. Kirton's regiment had Newcastle himself as its commander-in-chief – hence 'Newcastle's Whitecoats'.

At November's end Cavendish left Newcastle at the head of the 8,000-strong Northern Army, heading south with the aim of entering Yorkshire and securing York. Awaiting him on the south bank of the Tees was a small body of parliamentary horse and foot commanded by Captain John Hotham (1610–45). The riverbank at this point is a steep bluff. Hotham's detachment included about 250 infantry and two cannon, and it was up here, overlooking the bridge, that he put his artillery pieces and the musketeers. If musketeers concentrated fire on the bridge, the field pieces could aim into the main force that was held up on the other side. At first this seems to have worked; a Royalist sally onto the bridge was repulsed. However, Newcastle turned the tables by targeting the defenders with his own artillery. Outgunned, Hotham withdrew and the Northern Army marched into Yorkshire. Newcastle entered York two days later. We shall meet him again soon.

The action at Piercebridge was one among dozens that were fought along Dere Street and its continuations at different times. Among them was the halting of Scottish invasion at the Battle of the Standard (1138), the clash between Edward II and his barons at Boroughbridge (1322), and the brutal meeting of the Yorkist and Lancastrian armies at Towton on Palm Sunday 1461. Towton was fought close to the western margin of the Vale of York, on the limestone ridge.[12] There was a kind of inevitability about this. To the east, sluggish rivers flowed through marsh and fen, while Pennine hills rose to the west. The ridge was thus the only route for an army on the move between the east Midlands and the north, and its narrowness made it a pinch point. The road running along it was thus repeatedly contested in wartime and guarded by a string of castles in peace.[13] For the same reasons it was often a place of muster and stand-off, as between insurgents of the Pilgrimage of Grace and the army of the Crown near Doncaster in October 1536, or rebels and government during the Rising of the North in November 1569.[14]

North of Ripon the limestone outcrop disappears beneath a blanket of glacial drift that forms the gently undulating floor of the Vale of Mowbray. The Vale enjoys a sense of containment between the scarp of the North York Moors and the foothills of the Dales. In places it is only fifteen miles wide, and it was accordingly a kind of ginnel between northern and southern Britain. York's position between these two de-files, the limestone and the Vale, as well as its commanding position in relation to east–west communications, made it the key to the north. This is why at least ten instrumental battles were fought in the city's vicinity between the tenth and seventeenth centuries,[15] and it is why Newcastle moved to secure it for the Crown in December 1642.

Along their way the Northern Army passed places evoking older struggles. The first of them was barely an hour's march to the south. It is called Stanwick, just off Dere Street, where nearly five miles of monumental earthworks enclose an area larger than the City of London.[16] The Tudor antiquary John Leland wondered if this had been 'a camp of men of war' or the 'ruins of some old town'.[17] We now know that the complex evolved between the early first century BC and around AD 70, and that it may well have been the seat of Cartimandua, ruler of a northern upland people who were known to the Romans as the Brigantes.[18] According to Tacitus the Brigantes were at first pro-Roman. Objects excavated at Stanwick bear this out: in the mid-first century the Brigantian leadership was being showered with luxury goods brought from afar. However, internal dissent led to the ousting of Cartimandua and gave pretext for a full Roman takeover of Brigantian territories.[19] It was against this background that the legionary fortress was begun at York in the AD 70s, and Dere Street's miles of straightness were ruled across Yorkshire.

Eight miles further on, the Royalist army arrived at Catterick Bridge, where Dere Street crosses the River Swale. A few years after the subjugation of the Brigantes the Roman army put a fort here to guard the crossing. Civilian settlement and craft-working grew up on both sides of the river, and there may have been wharves for transhipment between road and river.[20] The place was called *Cataractonium*. Catterick links other times and places. A Welsh poem in a thirteenth-century manuscript contains funeral songs for members of an elite force who died in battle against English dynastic leaders at *Catraeth*, apparently around AD 600.[21] Catraeth was very likely Catterick or somewhere in its vicinity. The defeated force represented *Yr Hen Ogledd*, 'The 'Old North': the Welsh name for early medieval peoples in southern Scotland and northern England who were remembered as culturally and linguistically British in the face of English expansion. Wales became the keeper of memories of this identity, and behind Yorkshire is a greater Wales. The Old North was accordingly a fount of Welsh literature, homeland of poets like Taliesin and Aneirin (fl. sixth century), and a locale for figures like Arthur, who resolves as a regional strongman in the aftermath of the environmental crisis in the later 530s.[22] Connection between Wales and the north is more than memory: if you stand on the summit of Ingleborough you can see Llywelyn in Gwynedd, and from Snowdon on the clearest days you can look into Scotland.

Echoes of the Old North can be heard all along the Great North Road. Between Wetherby and Doncaster are place names with the affix '-in-Elmet'. Elmet was a British kingdom in the fifth and sixth centuries.[23] Its boundaries are obscure; like Deira it may have been more in the nature of a dynastic grouping than a territory with fixed limits, but its heartland seems to have coincided with much of western Yorkshire, and since places with names '-in-Elmet' occur along the limestone ridge it is likely that they trace its eastern march.[24] Later sources say that Elmet was ruled by British kings until the early seventh century, when the Northumbrian king Edwin (*c.* 586–633) expelled its ruler and annexed the region.[25]

It is tempting to see such conflict as binary opposition, but the sources suggest more subtle relationships. For instance, when the various kingdoms are first heard of they have British names. Young Northumbrian princes at risk from elder rivals sometimes sought the protection of British kings. Several English rulers married British princesses, and Welsh tradition asserts King Edwin of Northumbria was baptised by the son of a late sixth-century British king rather than by the Roman missionary

Paulinus.[26] When Edwin died in battle against an Anglo-Welsh alliance somewhere near Doncaster in 632, his English warlord opponent bore a British name – Penda.[27]

Thirty-three years later Penda met his own death in battle at the hands of the Northumbrian king Oswiu (*c.* 612–70). The battle concluded a long and mobile campaign, during most of which Penda and his allies, who included a Welsh king, had had the upper hand. According to Bede it was fought in November at a place called *Winwæd* where Osiwu's smaller force caught the Mercian army while it was apparently strung out crossing a river in flood.[28] We do not know where Winwæd was, but Bede says that it was in the district of *Loidis*, which puts it in the vicinity of places near Pontefract bearing names with the first element *Ledes* – like Ledsham, Ledston and Leeds. Ledsham and Ledston are on the ridge beside Higden's Ryknild Street.

A good case has been made for identifying Winwæd with a point on the River Went nearby.[29] A battle where the River Went carves through the limestone ridge on its way to the Humber marshes fits the narrative of pursuit, and would put Winwæd within a cluster of actions fought in close proximity at different times – among them, possibly, *Brunanburh* (937), where Æthelstan's defeat of an allied force of Scots, Irish and Danes made 'for the ultimate unity of England'.[30] Also within this area are earthwork enclosures either side of the Roman road, between which ran rock-cut ditches and banks. Parts of this system, at least, date from the late Iron Age. The fact that it was designed to control movement along the ridge implies that the road built by the Romans followed an older alignment.[31] How much older? A chain of ceremonial monuments runs beside Dere Street. Among them are earthen henges, parallel-sided avenues, and stone pillars. They date from the third millennium BC, and occur in localities where rivers flow out of the Pennine Dales into the Vales of Mowbray and York.[32] If people moved between them on a regular basis, their paths prefigured Dere Street by 3,000 years.

Battle names stir memories and imagination. Dere Street and Ryknild Street are thick with reminiscence and nostalgia for better days, with stories round the fire of Cartimandua, Catræth or blood flowing like water on Cowton Moor.[33] We can think of the Ninth Legion marching to oblivion in the second century,[34] or Yorkshire's gentry bringing out their old banners as they rode out to join the rising in 1569, or even of Robin Hood, whose earliest recorded stamping ground coincided with the area near Wentbridge where Penda met his end.[35] How much of this might have impinged on the poetical Newcastle as he marched towards

York in 1642 we cannot know, but he was heading towards the greatest battle of all.

By later summer 1643 Newcastle's force had wrested most of the West Riding from Parliamentary control and driven Ferdinando, Lord Fairfax, and his son Thomas into the refuge of Hull. Hull was heavily defended and permitted communication and supply by sea. The King demanded that the town be put under siege. Newcastle obliged, but from October was himself set back by a break-out led by Sir Thomas Fairfax and Oliver Cromwell. The Crown's hold on the north was further threatened by the arrival in Northumberland of the Scots Army of the Solemn League and Covenant. It was led by the veteran soldier Alexander Leslie, the Earl of Leven (1582–1661), and crossed the border in January 1644. The Scots' aim was to secure the port of Newcastle and its link with London, and to confront the northern Royalist army. To counter it Newcastle was obliged to split his force, taking part north and leaving the rest in Yorkshire under John Belasyse.

Limited engagements ensued, often in harsh weather. Leven's army crossed the Tyne upstream of Newcastle, edged south-eastwards and took Sunderland. Early in March, Newcastle withdrew to Durham, aiming to bring the Scots to battle at a place of his choosing. But there was no knock-out blow. Meanwhile, Royalist forces in Yorkshire came under attack from Thomas Fairfax and John Lambert in the West Riding, and ever-bolder cavalry sorties out of Hull. On 11 April, Belasyse and many of his infantry were captured in the course of a sharp Parliamentary assault on Selby. York, capital of the north, was now at immediate risk. Newcastle disengaged and with Leven on his heels force-marched down Dere Street to secure the city.

The armies of Leven and Lord Fairfax joined forces 'neare Weatherby' on 18 April.[36] Just to the east of Ingmanthorpe is an area called War Fields. In 1971 I sometimes walked this way to catch the bus to York. Close to the place where the footpath crossed Ainsty Beck there were traces of entrenchments and a rampart. Local opinion held that they were dug during the Civil War. If so, there are several episodes to which they could relate. However, it is likely that this was where Leven and Fairfax joined forces. The area is just off The Rudgate, the road down which Leven's army marched from Dere Street, and with local features bearing 'Ainsty' names it is a likely meeting place. Moreover, the Allies paused for several days, and it is quite possible that temporary earthworks were cast up as defence against nuisance raids from nearby Royalist garrisons at Pontefract and Cawood.

Newcastle warned the King that unless reinforcement was forthcoming 'we shall be distressed here very shortly'. Preparations to ready York against attack had been in progress for some time. Gates and posterns had been blocked with earth; suburban buildings were demolished to remove cover for attackers and provide clear fields of fire. Earthen outworks had been constructed, together with embanked fortifications for artillery known as sconces.

The besieging armies took up position on two sides of the city, Fairfax east of the Ouse and Foss, Leven and the Covenanters to the west. Pontoon bridges were formed from boats lashed together to allow communication across the Ouse downstream. Despite periodic anti-siege sorties from nearby Royalist garrisons, the Allies consolidated their positions.[37]

On 3 June the besiegers were joined by the army of the Eastern Association, fresh from securing Lincolnshire, under the Earl of Manchester. Manchester's force took up position on the north side of the city, above the confluence between the rivers Ouse and Foss. York was now surrounded. One news-book that week reported preparations for an onslaught involving fifty artillery pieces 'with full resolution to assault York on all sides'.[38] Artillery exchanges ensued, accompanied by besiegers' sallies to seize several sconces. Newcastle sought to parry the growing onslaught and play for time by inviting negotiation. The besiegers took a key sconce 'and other places of advantage' and 'daily played with their Ordnance into the City'.[39]

If the Blind Fiddler existed it was now that his services would have been in most demand. Of course, even if there was a passable Roman culvert it is unlikely that its outfall would have been far enough beyond the outworks to provide for undercover in–out access. This said, there could be some actual circumstance behind the legend: the Allies tried to undermine the city's defences at several points and it may be that the defenders used an ancient culvert as a place in which to station listeners.

Besiegers faced more than bullets. More Civil War soldiers perished through hardship than died in action. Sustained privation led to low resistance, turning otherwise minor ailments into killers. Tens of thousands perished from problems with simple causes, like louse-born fevers, dysentery from poor hygiene, or lack of clean water that drove men to drink from puddles. In 2007–8 archaeologists working in the inner York suburb of Fishergate – Fairfax's sector – came upon the remains of 113 men in one grave. The skeletons showed few signs of combat and their burial in one pit at one time indicates that they died within a few hours of each other. The men had been stripped and laid on their sides, nameless and unremembered; each packed against the next, like tinned anchovies.[40] Few discoveries say more about how the Civil War was fought.

On 16 June the Allies attempted a synchronised assault from different directions. The effort miscarried when an operation to blast a gap in the precinct wall of the former abbey of St Mary was mistimed, allowing Newcastle's men to fend off the ensuing incursion. A fortnight's stand-off followed the failed attack.

The attention of the Allied generals was now fixed on reports from the other side of the Pennines. Back in late April, Rupert, the King's 25-year-old nephew and commander of the Royalist cavalry, had attended a war council in Oxford. The plan they agreed was for the King to maintain his position in Oxford while Rupert's younger brother Maurice secured south-west England and Rupert himself went north to the aid of the Marquess of Newcastle. Since then Rupert had been active in Lancashire, recruiting, garnering supplies and securing a hinterland for his forthcoming mission through the successive capture of Stockport (25 May), Bolton (28 May, with much bloodshed), and Wigan (5 June). Liverpool fell on 11 June, thus providing the Crown with a port for supply and communication with Ireland (Fig. 15). In the interim, however, Parliamentary forces had been converging on Oxford. Their approach dislodged the King, who left the city early in the month and now moved from place to place to avoid direct confrontation. The King wrote to Rupert on 14 June urging him to use all his force 'to the relief of York', to 'beat the rebels army of both kingdoms', and then to join him at Worcester, where their armies could combine.

Rupert's force was now about 15,000 strong. The Allies called them 'the Rupertarians'. They set forth from Preston on 23 June, crossed into Yorkshire, and arrived in the Royalist stronghold at Skipton on the 26th. Here there was a pause 'to fix our arms', train recent recruits, and send a contact patrol ahead to alert Newcastle to their approach. Rupert did not wish to give the Allies time to organise; after two days he was on the move again, advancing down Wharfedale and over its watershed into Nidderdale, reaching the royal-held castle at Knaresborough on 30 June.

Leven, Manchester and Fairfax assumed that the Rupertarians would advance by the shortest route: east from Knaresborough to Skip Bridge, where Dere Street crosses the River Nidd, thence straight down Dere Street for the last seven miles to the city. To encourage them in such thinking Rupert sent forward a strong body of horse to pose as the

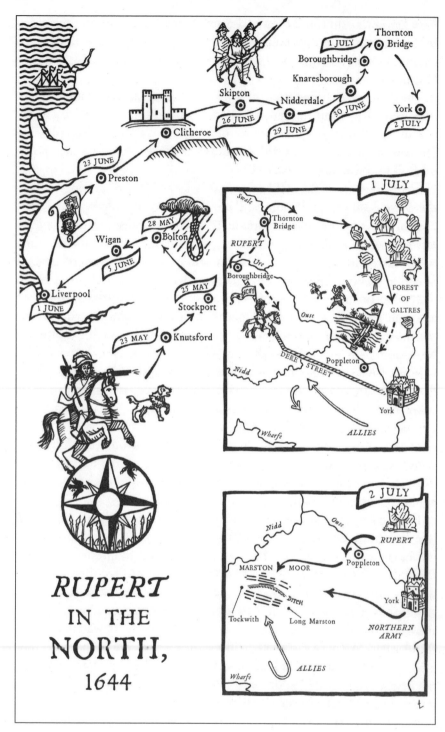

15. *Rupert's campaign in Lancashire and his approach to York, June 1644*

vanguard of an advancing formation. The Allies had already responded by lifting the siege and taking up position astride Dere Street on the eastern side of the Nidd. But as they waited on the morning of 1 July, the Royalist main force was eight and a half miles away, crossing the River Ure at Boroughbridge, whence it pushed on for another three miles to cross the River Swale at Thornton Bridge.

Rupert's army was now as far from York as it had been when it left Knaresborough, but south of Thornton Bridge the rivers Swale, Ure and Nidd flow into each other and become the River Ouse; hence, by reaching the east bank of the Swale, Rupert had opened the way for an unimpeded march to York, and the unification of his force with the Northern Army. The detour outflanked and outfoxed the Allies. Rupert rested his men for a few hours in Galtres, a former hunting forest north of the city, and sent word to Newcastle to bring his foot and horse to join him west of York just after sunrise the next day, Saturday, 2 July. Sunrise that day was at about twenty minutes to four. The rendezvous was to be in an area of The Ainsty called Marston Moor.

There is a feeling of difference about Marston Moor. There are villages on its edge, and York is only just down the road, but the district is its own place, wide-skied, low-lying, a blankness of rectilinear fields, dissected by drainage ditches and even now largely unsettled. In the seventeenth century much of it was rough pasture and tussock interspersed with heath and sedge. The area once formed part of a larger tract of lowland moor in which different parts bore individual names (like Low Moor, Scagglethorpe Moor, New Moor) or took names from adjoining townships and outside landholders (Hutton Moor, Moor Monkton, Abbey Moors). The whole is about three miles broad and roughly triangular, bordered on the south by a low ridge and by the rivers Nidd and Ouse to the north. Each river has its own manners. To the north-west, the Nidd wanders in a series of squiggles on its way to join the Ouse at Nun Monkton. When the Nidd is in winter flood this northern margin becomes a lake, frequented by honking geese and groups of visiting Whooper and Bewick's swans. Below Nun Monkton the Ouse broadens. At Poppleton, on York's outskirts, Leven's men and the Eastern Association had spanned the river with another floating bridge improvised from boards laid on a line of moored boats. On 1 July the bridge was guarded by a body of Eastern Association dragoons, who were driven off when Royalist cavalry arrived from the direction least expected. With the bridge in Royalist possession, the way was open for Rupert to march his army straight out onto the Moor.

In 1971 I passed Marston Moor every day on my way to and from York. There were two ways to make the journey. One began with a lift up the Rudgate to the remote branch line station at Cattal, followed by twenty minutes in a rattly DMU that smelled of diesel oil and pipe smoke. The other was by bus or car on the road from Wetherby. The road passed along the southern ridge, giving a long view north across the Moor, as if looking over the shoulders of the Allies towards the horizon of the Hambleton Hills.

The railway gave a Royalist perspective: the line between Hammerton and Poppleton runs alongside Dere Street, whence there is a view across the Moor towards the low green escarpment whereon the Allies formed up two miles to the south. The ceramic tile map of the North Eastern Railway even shows a station. It was opened in 1848 by the East and West Yorkshire Junction Railway, and although it closed to passengers in the 1950s the station house survives as a private dwelling. The alert passenger can glimpse a platform and the adjoining wooden signal box of 1910, still working, now listed, with its name plate announcing 'Marston Moor'.[41]

Whether or not the Battle of Marston Moor was the largest action ever fought on English soil, it was certainly one of the strangest. The action took place partly at night, two of the three victorious generals left the field before it was over in the belief that they had been defeated, and neither side knew who had won until hours after it was over. History knows that Marston Moor ended in Royalist failure, but the first report to reach the King told of resounding success.

The day began with difficulties and frustrations for both sides. When Allied leaders realised that they had been outmanoeuvred they decided to pull back to the safety of Cawood, where they would await reinforcement and keep hold of routes from Hull by which they were supplied. The Allied foot and artillery train accordingly set off towards Tadcaster early in the morning of 2 July, leaving 3,000 horse led by Cromwell and Thomas Fairfax on the southern ridge to keep lookout and protect their rear. By 9 a.m. units of Rupert's army were observed forming up on the Moor. The Allied main force was recalled, while the Parliamentary rearguard prepared to withstand probing efforts to push them off the ridge.

In York, meanwhile, there was discord. Lord Eythin, Newcastle's lieutenant general, and Rupert were not on good terms. Eythin and Newcastle challenged the case for an immediate full-on engagement. Intelligence in Newcastle's hands led him to think that the Allied federation of armies was on the point of breaking up, and that a pause to gain strength would enable the Royalist army to take fuller advantage at less cost. More practically, there were disciplinary difficulties with rank-and-file garrison members whose immediate priority was to plunder the trenches of their recent besiegers. In result, Newcastle did not arrive on the Moor until midday, and his infantry were not finally assembled until late afternoon. Rupert was thus unable to dislodge the Parliamentary rearguard from the ridge before the Allied main force returned. Rupert's heavy guns could not be carried across the floating bridge at Poppleton and there was insufficient time to bring them round through York, but some sixteen medium and light pieces were available and by 3 p.m. sporadic artillery exchanges had begun. Allied numbers amounted to perhaps 24,000, facing Rupert's force of about 17,500 horse and foot.

Towards five o'clock Rupert made an assessment of Allied intentions. Sunset that Saturday would be three and a half hours later. The Allied line had recently been moved forward a little, but although it remained formed up in battle array there was no sign of an imminent charge. It is possible that Rupert was unable to see the entire force – parts of it were in dead ground – and so underestimated its strength and readiness. At any rate, he decided that no charge was likely before the next morning, and since the last companies from York were still taking the field the remaining daylight could best be used by finalising preparations. In contrast, the Royalist dispositions were in full view of the Allied commanders, who by 7 p.m. knew that theirs was the larger force, and that their opponents were standing down. Following a conference with Manchester and Fairfax, Leven took the decision to attack. As one of Cromwell's colleagues wrote: 'About halfe an houre after seven o'clock at night, we seeing the enemy would not charge us, we resolved by the help of God, to charge them.'[42]

The villages of Tockwith and Long Marston lie to either side of the southern edge of the Moor. They are linked by a road that lay between the armies. Beside the road ran a ditch and hedge, behind which the Royalists posted musketeers in forward defence. Sometime after seven o'clock a volley of artillery gave the signal to advance. The Allies began to move. About a third of a mile lay between them and the skirmish line. This gave maybe four minutes for the first Allied elements to reach it and for the defenders to recover from their initial surprise.

On the Allied left, Cromwell's Ironside cavalry advanced steadily and crossed the ditch. After an initial encounter lasting maybe fifteen minutes the first line of Royalist horse was overwhelmed. However, the second line held; a prolonged cavalry engagement ensued, to which each side committed reserves. The Allies' greater weight eventually prevailed; Rupert was unhorsed and temporarily isolated. Cromwell was grazed by a bullet and left the field for the wound to be dressed.

To the right, meanwhile, events were unfolding in equal and opposite contrast. The eastern length of the ditch was deeper and wider than in the west. It presented a real obstacle, and when Allied horse led by Sir Thomas Fairfax arrived to cross it the defenders were ready. Fairfax's troops of horse met withering fire, and were in turn successfully charged by horse from the Royalist left led by Sir George Goring. The Allied reserve held out for a time, but eventually broke, so opening the way for Royalist assault on the Allied baggage train and a flank attack on infantry in the Allied centre. Fairfax and a few colleagues struggled through the battle for over a mile in search of Cromwell on the far wing.

It was growing dark. The battle was spread across a two-mile front and becoming increasingly chaotic. In the centre, the Allied foot had at first gained ground. However, in this kind of battle eventual advantage tended to lie with whichever side first brought up reserves and maintained sustained pressure through repeated assault. In this, the Scots foot were less experienced than their opponents. They lost momentum, and the know-how of Rupert's companies began to tell:[43]

'not only were many of Rupert's 'Irish' infantry in his first line more experienced than their opponents, but some at least of the seasoned northern foot in their second line were now advancing in support. Deployed as they were in chequer-board style formation, the northerners were able to move into the intervals between the divisions of the Royalist front line to join in the battle and launch a counterattack.[44]

Under this strain a number of units of Allied foot began to break up. Panic spread. The second line gave way. With rout now seeming imminent, Leven and Lord Fairfax lost composure and departed.

One Covenanter brigade stood firm. Its regiments were led by Viscount Maitland and the Earl of Crawford-Lindsay, both of whom withstood a succession of cavalry attacks and eventually captured their assailants' commander, Sir Charles Lucas, when his horse was shot from under him. In parallel, elements of remaining Scots regiments in the Allied reserve were reformed and the Earl of Manchester rallied remnants of several troops of horse and restored them to action. Royalist foot were dog-tired following their counter-attack, and when Allied units reformed and recovered their aggression they had little energy left with which to resist.

Cromwell knew nothing of the disaster on the far side of the battle or the crisis at the centre until Sir Thomas Fairfax arrived out of the uproar to tell him. What followed turned the action inside out. Cromwell led his Ironsides behind Rupert's original deployment and shattered the Royalist left from the rear. Having done so, he combined with Scots and Eastern Association foot to reduce what remained of Royalist opposition. Bullet scatters say something about the scale on which this happened. They are spread for nearly four miles. Either side of the road between Long Marston and Tockwith the dots of their mapped distributions occur in great drifts, like the arms of some great nebula.[45]

By 10 p.m. many areas of the battle were at a standstill. Groups of blown horse shifted listlessly, companies of pressed men threw down their weapons, and members of Lancashire foot thought of sweethearts and smallholdings and began to trek homeward. Their way led past Ingmanthorpe. Witnesses remembered these scenes silvered by a moon that was close to full. There was also nautical twilight: the battle was fought shortly after the summer solstice, when the northern sky was lit by green-blue light from the sun not yet far below the horizon. Among the last to stand and fight in this afterglow were two of the units that had been raised by Newcastle two years before. Led by Sir William Lambton and Col. Posthumous Kirton, Newcastle's Whitecoats formed a hedgehog of pike and fought on until only a few dozen were left.

Dawn came up on thousands of dead and injured men and horses lying out on the Moor, on footpaths, roads and under hedges for miles around. Further out, before and beyond Yorkshire's borders, some were still fleeing. Crows, buzzards and magpies feasted. Locals and Allied proxies stripped the dead, salvaged kit and shot just-still-living survivors to put them out of their wretchedness. Many Allied foot were already in a bad state, having slept in the open and been marched back and forth for days with little food and no clean water. Immediately after the battle some were reduced to drinking water puddled in the hollows of hoof-prints. Around midday Leven was briefed on the outcome and sheepishly returned to the site of the action that fourteen hours before he thought he had lost.

Newcastle, Eythin and Rupert were back in York in the early hours. There were recriminations.[46] Newcastle was distraught. He had given his fortune and years of his life to service of the King. The annihilation of the Northern Army was not just a military disaster but a personal humiliation. Later in the day he boarded his coach, ordered it to

Scarborough, and took ship for the continent and exile. Eythin and other officers of the Northern Army went with him.[47]

Rupert marched out next day with what force he had left.[48] For avoidance of the Allies they went first to Richmond, then back across the Pennines to Lancashire.

The siege of York resumed. Morale inside the city slumped; surrender talks opened. Lord Fairfax spoke for clemency. The Allies agreed that remaining defenders should be allowed to leave unhindered, and gave guarantees for the safety of the city. At the end of the campaign in the north, York survived. The cause of the King did not.

At Ingmanthorpe in deep turquoise twilight on a high-summer Saturday, listening to the Alpine's electric organ across the fields, it was not difficult to imagine the wallop of artillery, drum signals, or random sounds of different kinds of pain. But I never did. Nor did I ever wonder if a stir of wind or the movement of an owl might recollect some passing soul struggling to go home. But the perpetual hum of A1 traffic was a reminder of older journeys, of moments of national crux like Winwæd, the North Star clattering on its way to Edinburgh, the back-and-forth passage of armies along Dere Street, and the poetry of the Old North.

NORTH

Like the other Ridings, the North Riding existed before the Norman Conquest, and when first met in Domesday Book (1086) it was divided into thirteen wapentakes.[1] In 1662 it was given a lieutenancy and gained its own quarter sessions. The administrative county of the North Riding was established 1889, and was shortly afterwards divided into municipal boroughs, and urban and rural districts. Redcar, Thornaby and part of Stokesley were transferred to the new county borough of Teesside in 1968. The administrative county and lieutenancy were eliminated in the reorganisation of 1974, when much of the county's area was combined with York, the northern part of the former West Riding and a small part of the East Riding to form North Yorkshire.

The old North Riding contained some of Yorkshire's most distinctive and appealing landscapes: the heather-clad upland heath of the North York Moors, the Cleveland Hills, the undulating drumlin country of the Vale of Mowbray, waterfalls, and the hay meadows of upper Swaledale, aglow in June with sweet vernal grass, yellow rattle, the melancholy thistle, wood cranes bill, rough hawkbit and lady's mantles.

4

SPYALL, COWTONS AND THE
WOUNDS OF CHRIST

The church of St Mary, South Cowton, is not a building you are likely to find by chance. There is a sign, but it is small, faded and stuck in a hedge, and even if you see it the journey that follows down half a mile of potholed track is made with a growing sense that you are on someone else's land. But if you press on, past the sheds of an agricultural contractor, you arrive at a grassy lane that in turn brings you to a green space. And there, partly screened by a line of trees, is St Mary's. Beyond it, two fields away on a rise, is the late medieval house dressed as a castle where Sir George Bowes (1527–80) wrote to Thomas Radcliffe, the third Earl of Sussex, on a November Saturday in 1569.

> *Sir George Bowes to the Earl of Sussex, 8 November 1569*
> My bounden duty promised: pleaseth your good L[ord] to be advertised, the Earl of Northumberland is this day, with my lady his wife, passed by as it is said to Topcliffe, and meaneth to repair to your L[ordship]. He passed very quiet, with no great number, although there were very great speech . . . And now the rumours begin again to stay, which sure was very evil. I am this evening, or tomorrow morning, to pass to my house at Streatlam, where I intend to remain, ready to serve the Queen's Majesty, as I shall be commanded by your L[ordship]. And so humbly taketh my leave. From Cowton, the VIIIth of November 1569.[1]

Bowes was a career soldier and administrator. Since the accession of Elizabeth I his loyalty and ability had won him a reputation in government circles as a safe pair of hands, reflected in his appointment to the Council of the North in 1561 and as Yorkshire's sheriff two years later. His main home was at Streatlam near Barnard Castle, with other houses at Aske and South Cowton, where he now was.

Sussex was thirty-six miles away in the 'most famous and fair city

York', where he served the Queen as Lord President of the Council of the North. Writing from the Council's headquarters in the King's Manor, he replied the next day:

Earl of Sussex to Sir George Bowes, 9 November 1569
The Earl of Northumberland promiseth to come, but he writeth not when, and is yesternight come to Topcliffe. The Earl of Westm[orland] refuseth to come for fear of his enemies, except he should come with force, which would be cause of offence; and therefore I intend to write the Queen's commandment to them, for their repair to her Majesty presently. Now is the time to take principal heed of their doings, and therefore, good Sir George, have good spyall, and send good advertisement, and by unknown ways, for fear of intercepting; and if you see great cause, send many ways, for within six days we shall see the sequel to these matters. In haste, 9 November, 1569.[2]

Talk of rebellion had been heard around Yorkshire for months. Sussex's reports to the Queen had advised restraint. He had met the earls and thought their discontent would be moderated by winter.[3] Others agreed that the likelihood of a 'stir' would wane when 'nights were longer and colder and the ways worse and the waters bigger to stop their passages'.[4] The Queen remained apprehensive. Reports from her own agents said that some sort of upheaval was indeed in the offing, and that the earls of Northumberland and Westmorland were among those behind it. In mid-October she instructed Sussex to order them into her presence at court. Sussex initially moderated this by asking the earls to meet him in York. But the earls temporised. Sussex asked again; again they did not come. Finally:

Earl of Sussex to Sir George Bowes, 10 November 1569
Sir George – I have sent my letters to my L[ord] of Westmorland for commandment to repair to the Queen's Majesty, whereof I have yet no answer; and for that my L[ord] of Northumberland did dally with his confederates after he came to Topcliffe, I sent the like to him yesterday, who, showing himself much discontented in speech to my man, detained him until eleven in the night, before he could be horsed . . .

Next morning a band of horsemen arrived outside Northumberland's park at Topcliffe. Assuming that their purpose was his arrest, Northumberland 'conveyed himself away'.

The sheriff, in like sort, at midnight left his house. My lady
[Countess of Northumberland] sayeth there shall be no trouble, but
I will no more trust any words, and therefore pray you have good
spyall what is become of them, and advertise with speed. Some
think they be gone to Richmondshire. I have not yet heard from the
Court...

  At York, in haste,
    10 Nov. 1569, at midnight,
      Your assured friend,
        T. Sussex.[5]

The rumours of a plot were but the latest in a continuum of whispers
that had circulated since the Queen's accession eleven years before. In
1569, however, the whispering had grown. Members of the gentry and
nobility in Yorkshire and Durham discussed restoration of traditional
religion, either by ousting Elizabeth or by having Mary, her cousin,
declared as her successor. Among them were religious traditionalists
like Charles Neville, sixth Earl of Westmorland, and Thomas Percy,
the seventh Earl of Northumberland, for whom such talk raised un-
finished business. In 1537, when Thomas was eight, his father had been
executed for his part in the Pilgrimage of Grace. During the reign of
Mary I (1553–8) most of his family lands had been restored, and in 1557
he was reunited with key northern offices traditionally held within
his family. After Elizabeth's accession, however, Percy fortunes again
declined. Northumberland was obliged to quit the wardenship of the
Eastern March[6] in favour of Ralph Sadler, was rebuked by the Queen
for siding with her tenants in Richmond, and in 1566 was forced to yield
the rights to minerals on his own lands to the Crown. In the following year Northumberland became formally reconciled to the Catholic
Church. During the 'conduction' of Mary, Queen of Scots, from Carlisle
to Bolton Castle in Wensleydale in the spring of 1568 Elizabeth refused
to let Northumberland extend hospitality to her, instead assigning her
escort to George Bowes.[7]

Traditionalists seem to have felt their beliefs and interests to be im-
perilled. Among them were Richard Norton, 'Old Norton', a steadfast
Catholic who nonetheless had just served as Yorkshire's sheriff, his
brother-in-law Thomas Markenfield, and John Swinburne, a leading
servant of Westmorland. They in turn knew others with 'some substance,
status, and sense of grievance'. Even Sussex's own half-brother, Egre-
mont Radcliffe, was within their circle. There were reports of Elizabeth's

forthcoming excommunication, which in many minds would legitimise opposition to the sovereign.

The timing of the rising that now followed may well have been occasioned by the Queen's demand that the earls should present themselves to her. Believing that they now had nothing to lose, Percy and Neville declared themselves. In the next fortnight around 6,000 commoners rallied to their cause. Historians ask why they did so. The answer seems to be found in custom and memory, and as the officer bearing Sussex's message to George Bowes galloped back to South Cowton he was passing through a landscape thick with their signs.

South Cowton is in the Vale of Mowbray, among low hillocks and knolls that are combed through by flood plains of meandering rivers and tributary becks with names like Wiske, Howl and Stell. Place names in the Vale likewise have an individual music; Danby Wiske, Aiskew, Norton Conyers, Thirn, Hutton Hang, Smearholme, Thornton Watless, Middleton Quernhow and Potto could not be anywhere else. South Cowton is one of a cluster of Cowtons with horizons that are rimmed by the Pennines and the Cleveland Hills in a way that confers both intimacy and the feeling of a world beyond. There is a North Cowton and an East Cowton, which was also sometimes known as Great Cowton or Long Cowton. Then there are, or rather were, Atlow Cowton and Temple Cowton, both long gone but still remembered in names of farms and fields. The village of South Cowton has disappeared; its former street and house plots are now just creases in pasture, like grassy braille.

The Cowtons formed part of a medieval lordship – a mass of lands and manors under unified rule – that since the eleventh century had centred on Richmond, six miles to the east. *Richemont*, where Alf Myers had been held captive and under which slept Arthur and his knights, was a castle-town planted by the Normans near the mouth of Swaledale in the 1070s. The Lordship of Richmond covered the northern Yorkshire Dales. This was a huge area – greater than, say, today's Bedfordshire – and an older grouping of lands called Gillingshire lay behind it. Catterick and Gilling were among Gillingshire's key places, and as we have seen already they were places with a past.[8] Special waters flowed through Gilling. In April 1976 a boy playing near the bridge over Gilling Beck spotted a glint in the water. The lad went to investigate, reached down and pulled out a two-edged sword with a silver-mounted hilt. The sword had been made in the ninth century. The most likely explanation for this Arthurian moment is that this part of the beck was a traditional offering place – a point of passage between this world and another (Plate 8). Such places go back for millennia. With the coming of Christianity churches were sometimes put beside them, and they often occur by later bridges and causeways.[9] No surprise, then, that Gilling's parish church of St Agatha stands nearby, that it mothered a huge parish with many dependent chapels, or that one of the chapels was St Mary at South Cowton.

My first visit to St Mary's was on a drizzly December afternoon. I arrived in fading light, and found it sturdy, rather tall, and built of rough-coursed grey-brown sandstone rubble that was pierced by large fifteenth-century windows of matching design. The building had a sense of unity, as did the unusual way in which the nave aisles and two-storey porch rose to full height, and the chancel almost as far, so that the whole resembled a set of cuboid building blocks that someone had arranged in one go.[10] However, a closer look showed that this all-done-at-once look was achieved by cladding an older building with new work. Whoever staged this effect wanted you to think that you are looking at a complete rebuilding, but did not stretch to it and produced a makeover instead.

The instigator was not far to seek. The arms of Sir Richard Conyers (d. 1502), impaled with those of his first wife, Alice Wycliffe, are set in the west wall of the tower. Around the corner the message is repeated: over the porch entrance, and not to be missed by anyone going through it, are the arms of Sir Richard and Alice, together with an exhortation to pray for their souls.[11]

The door was locked. I rattled the closing ring in the hope that it was just stuck, but no, it was shut fast. In dimming daylight the chances of tracing the key-holder before nightfall seemed small. There was a priest's door in the south wall of the chancel, but any church explorer will tell you that such doors are hardly ever used and you never expect to find one unlocked – especially last thing on a December afternoon. Nonetheless, it was worth a go. I walked across, put my hand on the iron ring, and gave it a twist. The door opened. Just inside, amid a thin scatter of bat droppings, was the effigy of Sir Richard, carved full-size in alabaster (Fig. 16). His hair was shoulder length, lips slightly parted as if murmuring a prayer,[12] a high-cheek-boned face, blurred by five centuries of scratches and knocks, looking blankly up at the roof. The figure was portrayed in plate armour. A sword was slung from a sword-belt; at his feet, a lion, with one if its paws on the sword's tip. The hands were held in prayer; between them, a heart.

The effigy must once have lain atop a tomb chest, its sides no doubt panelled with niches occupied by coloured figures of mourners, relatives

16. Effigy of Sir Richard Norton, d. 1502, church of St Mary, South Cowton

and heraldic detail. Sometime in the past the tomb was destroyed, leaving the effigy to be propped on a couple of makeshift legs in the south-east corner of the chancel. Two alabaster effigies of women lay in the corresponding place against the north wall. They too had been displaced, one parked above the other on metal stanchions, as if on bunk beds. Their clothes and hair suggest that they were created around the same time as the effigy of Sir Richard, *c.* 1485–90.[13] One lady wears a headdress. She is probably Alice Wycliffe, the first of Sir Richard's two wives. The other has unbound hair and is often taken for his second wife, Katherine Bowes, who outlived him.[14] Very likely all three originally lay side by side. We know that their tomb was the focus of a chantry through which Richard endowed a chaplain to pray for his soul's safety as it journeyed through Purgatory, for the soul of Alice, and for his 'now wife' (Katherine), who would remain in charge of the manor until her death.[15] Watching over them was a stained-glass image of the Virgin coupled – again – with the arms of the Conyers.

Tombs like this were vivid, eye-catching and expensive. We do not know where the Conyers' tomb was originally placed, but almost certainly it was in the chancel and thus visible to all as well as close to the celebrant during Eucharist. In effect it turned the entire church into Sir Richard's mausoleum and Sir Richard's memory into a kind of permanent bystander at the sacrificial meal.[16] Villagers' worship thus became permeated by his memory. More than this, by a standard formula Richard's chantry priest was charged to pray for 'all Christian souls'. The result was a cycle of reciprocity in which all prayed for all, and parish communities that became visually centred as much on the remembrance of their gentry as the worship of God.[17]

The Conyers were the pre-eminent gentry family in fifteenth-century Richmondshire, seated five miles away at Hornby.[18] Richard was the third son of Christopher Conyers (died c. 1463), whose service to the Nevilles of Middleham and appointment as shire bailiff in 1436 began an association between the family and administration of the shire. Christopher's eldest son John (d. 1490) followed him as the Nevilles' senior retainer, and when Richard, Duke of Gloucester, became Middleham's lord in 1471, following Richard Neville's death at the Battle of Barnet, John was 'in a position to ensure a smooth transition of power'. In following months a number of John's relations were retained by the Duke. Among them were his son and grandson, his brother-in-law, William Burgh, his son-in-law, Thomas Markenfield, and Thomas Tunstall, his wife's half-brother.[19] Also on Richard's payroll was his own brother, Richard Conyers, who was appointed receiver of the lordship of Middleham, granted an annuity and given the former Neville manor of South Cowton.[20]

After Bosworth, Sir Richard and his brother were quickly reconciled to the new regime. Richard received an annuity from the new king as early as 1486.[21] Although he owed his land and station to the future Richard III, his effigy depicts him as Henry VII's man: he wears a Tudor livery collar embroidered with Lancastrian emblems.[22]

The tomb, the chantry, the remodelled church and the tower house were parts of a scheme by Sir Richard to identify himself and his descendants with South Cowton. Gentry liked to be associated with their own places. Such groundedness also created tensions: heads of families were loath to split their patrimony, and if younger sons received lands from their fathers it was usually on condition that the land would revert to the main heir when the beneficiary died.[23] The Duke of Gloucester's grant of South Cowton to Richard Conyers came with no such strings. The manor and its land, income and accompanying rights were placed in Sir Richard's hands to do with as he pleased. Here is the reason for the outburst of display: arms in conspicuous places, the tomb, the

17. Cowton Castle, in reality a fifteenth-century tower house, in 1907

church remade. It is almost as if he were trying to make up for lost time, and puts his house facing the church across the fields in a distinct light (Fig. 17).[24]

The house was described as 'new built' in 1487.[25] It belongs to a genre of castellated, towered mansions that were erected or modernised by Yorkshire gentry and nobility in the later fourteenth and fifteenth centuries. They are typically rectangular, of three storeys, with a taller tower at one or more of the corners. Examples built by retainers of the Nevilles and Duke of Gloucester can still be seen at Hornby (the Conyers family), Nappa Hall (the Metcalfes), Brough Hall near Catterick (Burgh), or Markenfield Hall outside Ripon (Markenfield). Spofforth (a seat of the Percys), Ripley (Ingleby) and Whorlton (Darcy, later Strangways) (Plate 9) are among a number that to this day are called castles. But defence was not their real purpose; they were embodiments of status and family pride.[26]

No son was born to Sir Richard. He named three daughters as co-heiresses: Margery, who had married Sir Ralph Bowes (*c.* 1468–1512); Margaret, who became the wife of Robert Danby of Yafford; and Eleanor, who married Robert Lassells of Sowerby, near Thirsk.[27] South Cowton passed to Margery Bowes, whose husband's family, seated just over the Tees at Streatlam, was later remembered as 'distinguished by civil or military talent in every successive generation'.[28] Margery and Ralph raised a daughter and four sons, the third of whom, Robert (c. 1496–1555), received a life interest in South Cowton upon Margery's death in 1524. Robert combined eventful careers as both lawyer and soldier, and when he died his interest in South Cowton passed to his elder brother Richard (1488–1558), who had married Elizabeth, daughter of Roger Aske of nearby Aske (more gentry, another tower house). George Bowes was their third born but first surviving son – which is why we find him in the tower house at South Cowton on 8 November 1569, writing to the Lord President of the Council of the North about the impending rebellion.

For several weeks the rebels had the government rattled. The rising began on Sunday, 14 November, when the Nevilles, Northumberland and Nortons rode into Durham, where some of them threw over the communion table, tore up an English-language Bible and celebrated mass. George Bowes reasonably concluded that 'they intend to make religion their ground'. Next day the rebels moved south to Darlington and Richmond. 'To get the favour of others,' wrote the chronicler Raphael Holinshed, 'they had a cross, with a banner of the five wounds borne after them.'

Everyone knew this banner: back in 1536 standards like it had been the emblem of rebellions we now remember collectively as the Pilgrimage of Grace. The Yorkshire-centred Pilgrimage was the largest and gravest in a series of semi-connected revolts through the autumn and winter of 1536–7, in the course of which a substantial proportion of males in northern England took up arms.[29] The stirs originated early in October in Lincolnshire, in reaction to a rumour that parish church goods were about to be seized by the Crown. Reaction in Yorkshire and other parts of the north followed; later in the month there was a stand-off at Doncaster between a royal force of perhaps 7,000 and an insurgent army estimated at 20,000–30,000, with more in reserve.[30] Following negotiation the Pilgrims disbanded, accepting undertakings upon which the King later reneged. One of the go-betweens in those negotiations was Sir George Bowes's uncle Robert, from South Cowton.

The Pilgrimage was a rising of the commons who were alarmed by the early Reformation and by rumours of confiscations and taxes. Its scale reflected a rare depth of alienation from a government and royal advisers perceived to be dictating policy from the south. At its height, northern gentry and nobility had lost their authority; for a time many of them perforce became assimilated to the commons' cause, wherein they worked from within to re-establish normality.[31] If the story of those upheavals was in part 'the story of how the understood leaders of society reacted when their authority was challenged by a sudden, widespread disturbance', how would the commons react in November 1569 when the understood leaders embarked on a cognate project under the same banner?[32]

Sir George Bowes moved to Barnard Castle, where he made his base with a small garrison and collated information flowing in from loyalists to forward to Sussex. 'The matter growth very hot,' he wrote on 17 November. He warned Sussex that the rebels 'draw away the hearts of the people'. Bowes was short of armour, weapons and funds ('the country of Yorkshire never goeth to war but for wages'). He asked for authorisation to draw on armour held at Newcastle and firearms from Berwick. The earls were now in Richmond, where they raised men 'first by fair speech, and after by offers of money, and lastly by threats of burning and spoiling'.[33] Intimidation was also reported from Northallerton, where civic leaders were warned by the Earl of Northumberland that if they did not provide one hundred men he would 'burn and spoil the town'. Houses of uncooperative Protestant gentry were entered and armour taken. Christopher Neville (about whom Bowes had reflected the day before: 'I wish he were further off') went into Cleveland 'to raise people'. By the 18th the rebels had arrived in Ripon, whence reconnaissance spoke worryingly of an increase in the number and quality of their horsemen, in contrast to their 'poor rascal footmen'.

By 19 November the earls had become 'a fiery flame burst out', barging into parish churches down the Vale of Mowbray to burn English service books and Bibles, break communion tables and set out 'proclamations and precepts'. Bowes noted that although the rising had begun under the leaders' own names, they were now projecting themselves as the Queen's true and faithful subjects, recruiting in her name, and using the official muster system and Norton's recent position as Yorkshire's sheriff to do so. The purpose of the rising was to counter 'crafty' advisers who 'abused the Queen, disordered the Realm, and now lastly seek and procure the destruction of the Nobility'.[34] A credibility contest was in progress. 'Daily people flee from these parts to the earls,' Bowes reported; '[I] know not what should be done to stay [them] for I have notified their unloyal and rebellious dealings, and with fair speech and bestowing of money, used those that came in the most gentle manner I could. But it availeth nothing, for they still steal after them.' On

22 November, Sussex wrote to Henry Clifford, the Earl of Cumberland, at Skipton, Sir John Forster and Henry Scrope (Wardens of the Middle and Western Marches, respectively) asking them to send 300–500 horse 'with speed'.

The rebels were 'hung in suspense' at Wetherby, fourteen miles west of York on the Great North Road, an area where we have seen that armies anciently mustered and moved to and fro.[35] Sussex's patrols watched for signs of an advance on York. He was jumpy: for several days work had been in progress to barricade the city: men of The Ainsty had been called up, river boats had been brought inside the defences, the Ouse ferry had been scuttled, residents of suburbs had been moved within the walls, and earth and rubble heaped up against gates and posterns. Sussex reasoned that if the rebels came as close as Tadcaster a clash would be inevitable; in this event he told Bowes to leave Barnard Castle 'and come hither, with all the force you can make'.[36] Sussex sent fifty pounds with the courier. Later in the day: 'have good spiall upon the earls, and, if they stay, where they be'. If horsemen arrived from the Wardens or the Earl of Cumberland, Bowes was to send them on to York with all haste. They should travel by an easterly route, 'by Newburgh, under the Moor side', to avoid a chance encounter with the rebels.[37]

The rebels dithered. After two days at Wetherby they shifted to nearby Knaresborough, then to Boroughbridge, and from there into the old Forest of Galtres. The forest offered a direct approach to York, where Sussex now waited with a much smaller force of 2,500 foot and 500 horse. But the rebels chose not to engage, though at each church they passed they 'pulled asunder the service books, paraphrases, and other books of scripture, translated into English'.

Meanwhile, an army was being assembled under government orders in the south, and Lord Hunsdon, Warden of the East March, and the veteran Tudor diplomat Sir Ralph Sadler had arrived in York. Elizabeth had had doubts about Sussex's performance and loyalty and part of Sadler's brief was to give an independent assessment. Sadler reassured her on both counts. He assessed rebel numbers as about 6,000 foot and 1,000 'well-appointed' horse.

Sussex still had a channel of communication with the earls, and it was perhaps through this that he and Hunsdon engaged in psychological operations, exaggerating both the strength of the Army of the South and the speed of its approach. This would explain why by 29 November the earls and their main force were back at Richmond, not far from where they had started a fortnight before.[38]

With immediate threat to York relaxed, it was Bowes and his garrison at Barnard Castle that now came under pressure. The rebels looked for his children, rustled his cattle, took his corn and meant 'to hold me shut with their horsemen'. In parallel they had taken the port of Hartlepool, through which they hoped to receive aid from King Philip of Spain. Bowes told Sussex: 'It is necessary that we be hastily relieved.' The castle was under fire and his ears were 'full of shooting'. He complained that if he had had the muskets or light cannon requested from Berwick he could 'beat them out of the town'. Bowes pointed out that Barnard Castle was only a day's journey from York, and that even the rumour of a relief force would probably scare the rebels off.[39] However, Sadler was unwilling to move until the force in York was 'furnished strongly with horsemen'. He told William Cecil, Elizabeth's Secretary of State: 'I do marvel much that my Lo[rd] of Cumberland, my Lo[rd] Scroop, and my Lo[ord] Wharton do lie still and do nothing, as far as I can hear.' Cumberland belatedly sent a hundred horsemen, but they were 'meanly horsed' and armed only with bows.

On 6 December, Sussex told Cecil that the situation at Barnard Castle had deteriorated. Sussex, Hunsdon and Sadler wrote jointly to the Queen:

> We hear the rebels do still besiege Sir George Bowes, and have yet done little, so as if he have victuals, we think they can not prevail; but he is so enclosed, partly by a river, and partly by them, as he can not advertise us of his state, nor we send to him.[40]

At the same time Sussex wrote to Admiral Lord Clinton, co-commander of the Army of the South, urging the rapid detachment of an advanced force: 'It would be a great pity such a gentleman should be lost.'[41] Next day Sussex attempted to pass a message to Bowes:

Earl of Sussex to Sir George Bowes, 7 December 1569
Good Sir George – I am sorry that I could not come sooner to your rescue, for, that, by the Queen's Majesty's special direction I have tarried for L[ord] of Warwick and my Lord Admiral, who will be at Boroughbridge by Saturday next, with 1200 horsemen, one thousand harqebusiers, one thousand armed pikes, and 2000 bills; and I, God willing, will be, the same night, at Topcliffe with 1200 horsemen, 500 harquebusiers, 500 pike, one thousand archers and 4000 billmen. I have also appointed the Earl of Cumberland, the L[ord] Scrope, the L[ord] Wharton, and Leonard Dacres, with all the horsemen and

footmen of Westmorland and Cumberland; and Sir John Foster, Sir Henry Percy, John a Selby, and the Treasurer of Berwick, with all the horsemen of Northumberland; 300 soldiers of Berwick; and Captain Carvell's band at Newcastle, now reckoned to be one hundred, to meet on Friday 11 December, to march forwards on that side towards you: so as, God willing, the one of these armies, or both, will be with you on Monday next, whereof I thought fit to give you knowledge.[42]

But it all took time. Sussex was still in York on the 11th and the Army of the South was not ready to fight until the following Monday, by when the action at Barnard Castle was over. The rebels had captured the outer ward three days before. The garrison was 'scanted' of bread, lacked water, and on the 11th many of its members began to desert by jumping off the walls.[43] Next day, soldiers guarding the gates abruptly threw them open and deserted to the rebels, 'whereupon Sir George Bowes, seeing the falsehood of his men, was driven to composition, and is with all his men, horses and armour that remained there, come away in safety'.[44]

The Queen's Council was told that Sir George had 'served very faithfully and stoutly', and that the rebels had 'spoiled him of all he hath'. Indeed they had. Bowes emerged from the siege to find his barns emptied, cattle taken and houses vandalised. When he reached his house at South Cowton he found the doors torn off, brewing vessels broken, kitchen chimneys wrecked, books mutilated and family records stolen. Even the windows had been shattered by the levering out of iron saddle bars. He told Cecil: 'I now possess nothing but my horse, armour and weapon.'[45] Sussex appointed Bowes as his deputy in Richmondshire and Provost Marshal of Durham and Yorkshire.

On 16 December, as the Army of the South arrived on the Tees, the earls advised their rank and file 'to make shift for themselves' and fled into Scotland. In the following months Hundson, Forster and Sussex led ferocious cross-border raids in pursuit of escapees and Marian devotees, many of whom went into exile on the continent. Northumberland was detained by the borderer with whom he sought refuge and sold to the Scottish regent. His wife, Anne, and Elizabeth I bid for his return, an auction eventually won by the Queen, into whose power Percy was delivered for execution at York in August 1572. His head was displayed on Micklegate Bar, where it rotted, the rest of him being buried unmarked in the since-demolished parish church of St Crux, Pavement. Westmorland escaped to the continent, where he engaged in further

Walter Mitty-like Catholic liberationist projects until his death in 1601.

In the last days of 1569 Sussex began to interrogate local officials to discover who had been the chief instruments 'to raise the people in this rebellion'. Bowes found that the horsemen had been best equipped and organised. They consisted mostly of gentlemen, their household servants, tenants and head husbandmen. Historians once supposed that the rebel army as a whole was assembled like this. However, it is now clear that less than a fifth of the rank and file were tenants or retainers of the earls.[46] Of those who can be categorised, over 80 per cent were yeomen who came from 'broadly supportive communities' together with smaller numbers of tradesmen and husbandmen. Out of forty-eight known rebels from Richmond, for instance, 'at least sixteen had been burgesses, school governors, and men with similar local responsibilities'.[47] Whatever air they were breathing, it was not the atmosphere of feudalism. While the northern rebellion was instigated by the earls, the many who rallied to it saw its cause as something that touched them directly. Or as Bowes advised Cecil after it was done: 'I am not able to give counsel; but, good Mr Secretary, advise her Majesty to look well to herself, and not think all gold that glisters: for it will fall out to be the greatest conspiracy that hath been in this realm this hundred year.' Bowes believed that many more would have joined the rising 'if it had not burst out before the time'.[48]

On New Year's Day 1570 Bowes sent Sussex a list of the gentlemen detained in Durham castle. Many of the names are those of families we met a century before: Markenfield, Norton, Metcalfe ... Conyers. Hangings of 'meaner' rebels began the following Thursday and continued for weeks. It was emerging that village constables had played a significant role in recruitment – another example of the way in which the rebels had harnessed the machinery of civil administration to their cause. The constables were now to be hanged in their own towns. Sussex told Cecil:

> Besides the execution done in the great towns, there shall be no
> town where any men went out of the town to serve the Earls, and
> continued after the pardon proclaimed; but one man, or more, as the
> bigness of the town is, shall be executed for example, in the principal
> place of that town.
>   The common people were dispersed when th[e] Earls left
> Durham; and, therefore, the execution is the longer in doing, by
> reason of the apprehending and examining of the constables;

otherwise the guilty might escape and the unguilty suffer, and none of the constables that be found faulty be spared.[49]

The Bishop of Durham advised Cecil that administration of justice was difficult because no jury could be assembled with members who had not been involved. Unless individuals were acquitted or pardoned 'the number of offenders is so great, that few innocent are left to try the guilty'.[50]

Late in January, George Bowes told his brother that since his appointment as marshal he had had 'scarce five hours of the 24 to rest my wearied head and bones'. Bowes described his progress through Durham, Richmondshire, Allertonshire, Cleveland and Ripon, and so to Wetherby to 'sift' rebels by martial law:

> In which circuit and journey, there is of them executed, six hundred
> and odd; so that now the authors of this rebellion is cursed of every
> side; and sure the people [are] in marvellous fear, so that I trust
> there shall never such thing happen in these parts again.[51]

While men dangled from trees and improvised scaffolds between Doncaster and Durham, misconduct by members of the Crown's army added to sorrows and piled-up resentments. On New Year's Day 1570 Sussex penned a furious protest to William Cecil about greedy officers who daily rode about the country 'seizing, and spoiling, and ransacking at their pleasures'. Cattle had been driven off and ordinary people ransomed 'in such miserable sport . . . as the like, I think, was never heard of, putting no difference between the good and the bad'. Sussex's orders to the sheriffs to protect the rebels' goods and cattle for the Queen were ignored. He was particularly incensed by Lord Willoughby, who roamed Richmondshire, 'taketh all and leaveth nothing, and driveth more cattle for his own portion than all the grounds he hath will feed . . . by which means her Majesty hath lost above 10,000 pounds'.[52]

If we take the earls and their supporters at their word, they rose to restore 'the true and Catholic religion', to thwart reformation by strangers, and from a kind of melancholic longing for a social order of former days. Nationally, their defeat created new martyrs, boosted English Catholic communities in exile, nurtured more plots, provided new resources for royal patronage, led to tightened control over the character of clergy, and fed anti-papal rhetoric for years to come. Above all, for Elizabethan loyalists it drove home the need for uncompromising clarity in the religious settlement.[53]

In Yorkshire, assets changed hands. Around sixty of the wealthier rebels forfeited estates, houses and property. A number of old families were removed from their traditional seats, newcomers moved into old halls and tenants found themselves under different landlords (Plate 11). The dynastic truncation has sometimes been seen as a line of cleavage. In practice, few of the wealthy gentry lost their lives. Some bought pardons for cash, while in a number of cases the Queen temporarily gained control of estates that reverted to the families after the rebels' deaths. Wives and widows of attainted rebels fared badly, while even the land of rebels who obtained pardons passed out of their wives' hands until the men died. The Queen took income from the Earl of Northumberland's lands for a number of years before entailing most of them to his brother.[54]

The effects of spoiling by the rebels and the Queen's own soldiers were felt for years.[55] Perhaps the greatest disconnection, however, was something inner, springing from the realisation that the public disavowal of things in which people like Richard Norton and Thomas Markenfield believed was going to be irreversible. The rebels had set out to restore 'ancient religion' in which prayers and masses for the dead linked them with the living, and the future of a place with its past. Local dynasties had flourished in the midst of their communities through prayers in their chantries and the sight of their effigies in 'village Westminster Abbeys'.[56] Their world of customs and rituals was one in which everyone had had a part. It was a world richly coloured and textured by images, lights, sounds and movement, and after the rising there was no going back to it. Or as the twenty-second of the Thirty-Nine Articles put it, doctrine concerning prayers for the dead, the adoration of images and invocation of saints was 'a fond thing vainly invented'.

An object that evokes both the old faith and the new, and the last split between them, comes from the church of St Martin, Womersley, near Doncaster. It is the head of a cross made to be carried in processions. The cross-head is formed from a wooden substrate surfaced with copper alloy plates and adorned with glass and enamel insets. At its centre is an image of the crucified Christ. To either side are fixings for lost flanking figures, presumably Mary and John. The cross was made in Limoges in the thirteenth century, and found beneath the chancel floor during alterations in the 1870s. It belongs to a class of devotional objects – like altar slabs, relics, carved images, churchyard cross-heads – that were sequestered in and around churches during the reigns of

Edward VI and Elizabeth I. The manner of their disposal, gently laid to rest, not smashed or thrown out, points to a mood of acquiescent farewell.

To this day, among the Cowtons, where the music of place names hums between the Wiske, Howl and Stell, it sounds in a minor key.

5

NO PLACE LIKE HOME: LACKAWANNA, NASSAGAWEYA AND CASTLE DISMAL

―――――

Wiske, Howl, Stell, Durnbach ... Durnbach? Durnbach is near Bad Tolz, about thirty miles south of Munich. To find its link with Yorkshire we must first look west, to Canada.

I was five, going on six, when I was taken to Canada. The Atlantic crossing in 1953 was the greatest event in my life so far – until we arrived at my grandmother's cottage on the shore of Point Pelee. Point Pelee (Canadians say it 'Pee-lee') is a triangle of south-western Ontario that juts into Lake Erie. Yorkshire people have an eye for a peninsula, perhaps because Spurn Point is one of their county's most distinctive features. But Pelee and Spurn are different. Pelee is twice as long as Spurn, and whereas Spurn is slim and curls to a spoon-like tip, a bit like the beak of some wading, mud-probing bird, Point Pelee resembles the blade of a dagger, broad at the base and narrowing to an apex so sharp that if you walk right to the end you can plant your legs to either side.

But to go back to the cottage: it was framed of timber, walled with narrow planks, and covered by roofing felt. There was nothing special about it, save that the forty-second parallel ran more or less through the living room. No, the thrill, the glorious, knock-down thrill of Nana Wearne's home was where it was: shaded by trees, on the crest of a beach, with Lake Erie at the beach's foot. And since Erie is the world's eleventh largest lake, in the eyes of a five-year-old it was the sea, because you couldn't see the other side, there were proper waves, and ocean-going ships crossed its horizon. Skinningrove's beach was stony, draped with ribbons of slimy brown weed and usually misted by drizzle. Here, you stepped out of the back door onto sand and the water was warm. It was all I could ever have wished for.

Amid the hum of cicadas, watching a neighbour called Nancy chugging out every morning in a maroon-coloured boat to check trap nets, two memories drifted back from movies we had seen on the boat. One was *Lilli*, which had a catchy song ('Hi Lilli, hi Lilli, hi-lo') that I liked, but a story involving four puppets whose knowing stares struck me as a strangely unsettling. The other was the tornado sequence in *The Wizard of Oz*. Dorothy's wooden Kansas farmhouse reminded me a little of Nana Wearne's cottage, and the risk that we and it might be sucked into the sky became palpable one afternoon when Nana Wearne said that tornado warnings had been given on the radio. Towards evening the sky darkened and the wind rose. I stood on the beach, marvelling at the swelling waves and the lilac and gold-tinted flickers of electric storms in the direction of Ohio. The storm grew. I moved into the lee of the cottage, only to be thwacked on the back of the head by the frame of an unfastened mosquito screen that swung out in a gust of wind. A sense of merging with the film was now strong. I ran indoors. Nana Wearne was matter-of-fact: if a tornado came, we were to leave the cottage and hunker down in the ditch across the road. Eventually I fell asleep, lulled by roaring wind and the creak of heaving woodwork, and woke next morning to the gentle knock-knock-knock of Nancy's diesel as her fishing boat headed out across a glittering calm.

Sixty years passed before it crossed my mind to check Ontario's weather records for those days. In doing so I found that our arrival at Quebec had coincided with one of the greatest natural catastrophes in twentieth-century American history. Tornadoes among the Great Lakes are unusual, but on 7 June, when our liner was in mid-Atlantic and I was watching *The Wizard of Oz*, disturbed weather began to stir in Iowa, Nebraska and Wisconsin. The next three days saw the birth of a succession of tornado families, many of which tracked east across the Great Lakes and on into New Hampshire and Massachusetts. By the time we docked, 245 people had been killed and hospitals were crammed with thousands of injured. Almost half the casualties were victims of a single tornado that had whirled through the city of Flint in nearby Michigan, hoovering up houses, flinging cars for hundreds of yards and flaying the bark from trees. Naturally, nobody told me about this; when the new alert was given a few days later Nana Wearne's concern was that history might be about to repeat itself.

Nana Wearne was then in her seventies, smallish, spry, alert, white-haired, with a narrowish face and a ready smile. The smile reminded me of Aunt Hattie, her sister, back in Carlin How. So did her speech. She used strange words like dishcloth (tea towel), cookie (biscuit) and gas (petrol), but her dialect was of Yorkshire, when afternoon was 'af-ternearn', break was 'brak', 'yance' meant once, and 'sin' was 'ago'. Point

Pelee is famous for its savannah habitats and the range of its birdlife, but alongside the Boreal Chickadee and the wondrous annual migration of Monarch butterflies, here was a living link to God's Own County three and a half thousand miles away.

For years, I supposed that the Wearnes' journey to Canada in 1925 was the result of some isolated impulse, a step they took on their own. It all seemed very daring: I imagined Nana Wearne and her husband, Tom, discussing whether it would best to stay or go, and wondered how they measured their dreams against the life they already knew. Of course, they must have known others who had similar plans, and they were exposed to government advertising and agents who toured Britain to attract migrants with skills that Canada needed. During the 1890s and 1900s there was daily encouragement to migrate through advertisements in Yorkshire newspapers like the *Hull Packet* and *Leeds Mercury*. Amid legal announcements and small ads for Owbridge's Lung Tonic, Lockyer's Sulphur Hair Restorer or Taraxacum ('a vegetable fluid combination' for the cure of 'General Stomach Derangement') the Wearnes had been reading about cheap emigration for years. At the end of the nineteenth century shipping lines advertising in Yorkshire papers emphasised the performance of their vessels. The Beaver Line promised 'magnificent, full-powered, fast Steamers' and a 'Liberal supply of food' during the voyage. The 'full-powered Iron Steamers' of the American Line would whisk you from Liverpool to Philadelphia, while advertising by the Dominion Line (which had 'Eleven Powerful Steamers') drew attention to the government-assisted passage to Canada for agricultural labourers and domestic servants. The Allan Line specialised in providing advice for prospective emigrants to 'The country for agriculturalists', while after the Great War the Empire Settlement Act 1922 provided new inducements to move.[1] This was only part of the picture. Census returns, family records and passenger lists reveal Emily and Tom not as lone decision-takers but as grains in a vast flow which in one way or another involved nearly everyone they knew.

To go back a step: the lady in the large hat in the beach photographer's car at Scarborough in 1910 was Hannah Smith, Nana Wearne's mother. Hannah's father was Richard Coleman, who had been a poor child of the Northamptonshire parish in which he was born. He became a railway labourer and married Elizabeth Walker while he was a member of the 2,300-strong workforce that dug the 2.14-mile Bramhope railway tunnel north of Leeds in 1845–9.[2] (So many miners died during its excavation that a monument imitating one of its portals was

18. Memorial in Otley churchyard to twenty-four men who died during construction of the Bramhope tunnel, 1845–9

afterwards erected in Otley churchyard to remember them (Fig. 18). After Bramhope, Richard and Elizabeth moved to Marske, just up the coast from Skinningrove. They had five children. The eldest of them was Annie, born in 1851, who married an ironstone miner called Frank Nichols when she was seventeen. Over the next twenty years eight children were born to them in Marske, and when Annie's mother died her father came to lodge with them. Then, in 1888, Annie and Frank moved the family to Pennsylvania. They settled in Oliphant, a township in the newly created county of Lackawanna. Like Skinningrove and Redcar, Lackawanna specialised in iron and steel, and drew on Pennsylvania's plentiful supply of anthracite to do so.

Next to go was Nana Wearne's brother-in-law, Charles. Charles was a carpenter. He sailed for New York in the spring of 1905. He too set up house in Lackawanna, and in due course one of his sisters settled there too. Back in Yorkshire, meanwhile, Annie Nichols's younger sister Hannah had married Jack Smith, known to friends as 'Wild Jack' on account of his profuse red hair. Like his father-in-law, Richard Coleman, Jack was a labourer; during the first years of their marriage they lived not far from his birthplace in Northamptonshire. However, in

1885, in the depth of the agricultural depression, Jack abandoned farm work for mining and moved his family to Skinningrove. A number of his siblings and children then followed Annie and Charles across the Atlantic. Jack's sister Eliza married an insurance agent called Fred; they sailed for Canada in 1913, two among 182,500 British people who made the journey that year.

Thomas Larke, a blast furnace charger in the ironworks at Carlin How, married Nana Wearne's younger sister Ethel in 1906. He arrived in Canada in 1920 and found work in Windsor, Ontario. Ethel's brother Jack made the crossing two years later. He too went to Windsor and so did Tom Wearne, my grandfather, when he followed in 1924. Tom lodged with the Larkes while he looked for work and found a house into which his family moved when they arrived in the following year. The sale of household goods before they left stirred no emotion until a family arrived to look at their piano. Tom and several of the children had played on it year in, year out. The prospective purchasers asked Tom's eldest daughter to play something, so they could hear it. She did so, then wept.

In tandem with official encouragement and inducements, then, ran informal networks of relatives and friends. Through them, families back in Yorkshire were kept abreast of life in Canada – like the electric sewing machine and vacuum cleaner which lightened the drudgery of Nana Wearne's housework after her arrival in 1925. If they decided to follow they knew what to expect, and they were supported when they arrived.

Such networks had been forming since the early eighteenth century. The numbers of those who left had fluctuated. During the 1820s, for instance, fewer than 17,000 people a year had emigrated, whereas in 1831 the number was 83,000, the majority going to British North America and the United States. Emigration reached a nineteenth-century peak in 1852, when 368,000 departed, the majority to the USA. By this time, Liverpool was the primary port of departure. It became so because it owned a large tonnage of merchant shipping, its hinterland adjoined two main areas whence emigrants came,[3] it was accessible by railway, a local infrastructure of agents and government officials had grown up, and it faced west.[4] Between 1860 and 1900, 86 per cent of the 5.5 million people who left the British Isles passed through Liverpool.[5]

Earlier in the century smaller numbers of emigrants had embarked from a larger number of ports. In 1830 there were some nineteen ports of departure around Britain, of which two – Whitby and Hull – served Yorkshire, or three if you include the inland port of Stockton-on-Tees just across the border in County Durham. In that year five vessels carrying emigrants left Whitby. The largest of them was the newly built *Gulnare*, of 338 tons, which sailed on her maiden voyage with 230 emigrants. The *Jackson*, in contrast, carried just thirty-five, and the total of 405 passengers who passed through Whitby in 1830 was small in comparison with the 2,592 who left from Hull. The traffic from Whitby and Hull was not the specialised movement of people that it later became. It was in part a result of the resourcefulness of the timber-importing ship-owners, who early in the nineteenth century realised that paying passengers made better sense than ballast on the outbound voyage.[6]

The average tonnage of the vessels in which Yorkshire migrants travelled in the 1830s was 300. Before the age of steamships, a normal voyage took from four to six weeks, and of many risky things that could happen during that time disease was among the worst. Passenger numbers were sometimes thinned by typhoid and cholera; infectious ailments like measles spread swiftly. Some ships' masters were unscrupulous, as on the *Captain Ross* out of Whitby in 1834, where the water provided was so foul that passengers were forced to drink ale or porter that could only be bought from the captain.[7]

A question, then, is what it was that caused families not only to move to another part of the planet, but also to put themselves and their children in harm's way while doing so. The question has no one answer, but links rather to a web of influences which operated variably as time passed. Aback of them were forces of both push and pull.[8] For much of the nineteenth century push was the stronger of the two, reflected in the government view that emigration was a way to reduce poverty and unemployment,[9] and a working-class perception that Canada offered means for self-betterment that would not be found in Britain.

The period was one when population grew faster than did opportunities for employment – a combination that put pressure on poor relief and held down agricultural wages.[10] The mismatch between workforce and employment was aggravated by the introduction of farm machinery and latterly, in some areas, by the conversion of arable to pasture, both of which reduced requirements for labour. Farmers themselves often faced hardship and sometimes crisis. In grain-growing areas the absorption of lesser arable holdings by larger ones could marginalise small farmers, while imports of foodstuffs in the 1880s and '90s undercut local prices and contributed to farm failures.[11] Many farms were too small to remain viable in times of strain, and the custom that only eldest sons should inherit them deprived younger brothers in large families of local prospects.

Of course, much migration was internal: in industrialising economies towns provide an increasing proportion of employment and so draw in poorer agricultural workers from the countryside.[12] But towns, too, and entire industries within them, were subject to local and regional depressions caused by fluctuations in demand and prices, external competition, and new machines that took over tasks once done by people.

These are examples, and they make the push behind emigration look like dry economics. But other influences were important. One of them was imagination: for some people, and at certain times, dreams and hopes came to the fore. Another was religious freedom. Dissenters – Methodists, Baptists, Quakers, Catholics, Unitarians – resented paying tithes, rents or rates to a Church of England to which they did not belong and in whose churchyards they did not wish eventually to lie. Rifts within communities could arise from injustice, as in the parish of Danby on the North York Moors, where – as in most places – there was an undertow of local irritants.[13] Among them was a clash of interests in 1862 between the lord of the manor who wished to manage thousands of acres of heather moorland for sport and freehold commoners who by custom dug the underlying peat for fuel to keep their families warm.[14] The cumulative effect of individually minor economic stresses and grievances could be large: the 1881 census shows that of 1,862 people then living in the United Kingdom who had been born in Danby and its neighbouring parish of Westerdale, 1,133 – nearly 61 per cent – had left.[15]

A succession of schemes was organised to promote emigration to Canada. Some involved a direct connection between a part of Canada and parts of Yorkshire, as in the 1770s when the Lieutenant Governor of Nova Scotia recruited 900 citizens of the North and East Ridings to settle in the Chignecto Isthmus. The Governor, a leading merchant in Halifax, had inside knowledge of tenants' grievances in parts of Yorkshire where enclosure and rent increases had given rise to discontent. Many of them were Methodists, and the prime sites that were offered to them came with a guarantee that 'there are no game-laws, taxes on lands, or tithes in this province'.[16] Some travelled as families, others went individually, inviting relatives or friends to join them when they had become established. James Metcalfe, for instance, was one of sixty-two emigrants from the North Riding aboard the *Duke of York* when she set sail from Liverpool in 1772. Metcalfe came from Hawnby, a small village in the North York Moors. He bought 207 acres by the Maccan River in Nova Scotia, farmed successfully, and wrote to his fiancée, Ann

Gill, near Easingwold, urging her to join him: 'may ye Lord bless you and conduct you safely hither'. The letter took two years to reach her. Ann set off. They were married the day after her arrival.[17]

After the Napoleonic Wars a government scheme to encourage group emigration to Canada offered assistance of ten pounds. During this period settlers from Yorkshire spread into south-western Quebec and then into Upper Canada (later Southern Ontario), where they became concentrated along the north-west shore of Lake Ontario and in the south-western peninsula[18] – the area where Tom Wearne found enclaves of Canadians with Yorkshire forebears when he arrived nearly a century later. Between 1819 and the mid-1820s an advertising drive in north-east England invited settlers to southern New Brunswick, resulting in clusters of Yorkshire families in places such as Moncton, Petilcodiac and the Yorkshire-named Coverdale. Some families set off in groups, as from Bewholme in the East Riding in 1817 to join a Yorkshire community in Queen's County. There were similar Yorkshire parties at Gagetown (from 1824) and Annapolis Royal in Nova Scotia. A Colonial Land and Emigration Commission was established in 1833, offering free passage to agricultural labourers or shepherds under forty, to females with experience as farm servants, and in particular to young couples who had not yet had children. In the following year, the Poor Law Amendment Act included a provision to apply funds from parish rates to support the export of paupers. By this time localised patterns of origin had begun to dissolve.

The strongest and most sustained campaign arose out of Canada's prairie development policy. By the 1870s eastern Canada was becoming urbanised, whereas the west remained undeveloped. Canada's government reckoned that by populating the prairies with family farmers from Britain and the United States they would support the industrialising cities to the east.[19] To do this the government embarked on a sustained drive to attract farmers and agricultural labourers. One of its tools was the offer of a free 160-acre farm.[20] Another was the transcontinental railway, built between 1881 and 1885, which assisted settlement and carried its fruits. A third was *Canada West*, a magazine which depicted prairie life and projected images of a modern society that revolved around family farms. On the magazine's covers were pretty wives, happy children, golden sheaves and livestock thriving under wide blue skies, depicted in conjunction with messages about light taxes, free schools and healthy climate. Prairie Canada was 'the new El Dorado', a 'new homeland' of 'rich virgin soil' where there was 'nothing to fear' and there were 'homes for everybody'.[21] The idea of plenty was a thread that ran through the campaign; a highlight of the Canadian Pavilion at the British Empire Exhibition held at Wembley in 1924 was a sculpture of the Prince of Wales and his horse carved out of Canadian butter.[22]

Back in Yorkshire, families had long been able to test official inducements against the real experiences of relatives and neighbours. In 1867, for instance, a fourteen-year-old, Elizabeth Palliser, wrote from Potosi, across the border in Wisconsin, to her uncle and aunt, who farmed at Darley in Nidderdale. Elizabeth listed family members who were poorly, recovering, well, dead or about to marry, and reported on the purchase of an adjoining seventy-acre farm for 1,200 dollars, bought 'very cheap as the man wished to go to another state'. The recent Civil War had been expensive for them because their county had levied a local tax to hire stand-ins for young farmworkers who otherwise would have been liable to the draft. Then the farm:

> We raise sugar cane and make our own treacle from it. Last fall we made about one hundred gallons. Mother says it is sweeter and better than any she ever got in England. Now I must tell you about our stock: we have 17 head of cattle young and old together with 40 head of sheep and five horses young and old. This year we fattened 18 hog and we have 21 for next year besides 2 or 3 sows. We have lots of ducks and geese and chickens. We have plenty of fine apples we over one hundred apple trees and about half of them bear every year and we have some pear trees but they are not large enough to bear any yet, we can go in to the woods and get as many plums and gooseberries as we want but we have plenty on the place besides plenty of currants.[23]

Place names were transferred from Yorkshire to Canada.[24] Whitby was one of a number of townships in Ontario that recollected Yorkshire. Others included Scarborough, Malton, Hornby, Pickering, Sheffield, Bradford and Rosedale. Scarborough illustrates the kind of casual process by which a name could be transferred. Elizabeth Simcoe, wife of the Lieutenant Governor of Upper Canada (1791–6) bestowed it because Lake Ontario's shoreline here reminded her of Yorkshire. Toronto itself was briefly called York, though after Prince Frederick, Duke of York (the second son of George III), rather than directly after the city. York-Toronto has a Don River, a Humber, even a Trent, recalling its confluence with the Ouse to make the Humber. But this was no remade toponymy. In among the Yorkshire names were names from all over England (Tiverton, Cheltenham, Islington, Aldershot,

Watford, Battersea), other parts of Europe (Sligo, Elba), from settlers (Orangeville, Flesherton) and aboriginal peoples. It is as if different atlases had been shredded and handfuls of their disembodied fragments intercalated at random.

There was nothing of old Yorkshire about the shaping of the places themselves, no nomadic boundaries wandering between becks, legendary burial mounds and isolated trees. Canada was parcelled in squares and rectangles with an unswerving rationality that ran for hundreds of miles at a time. An example is the area around Malton, which came into British hands in 1818 through the purchase of a thousand square miles of uncleared land from indigenous Mississauga people. The laying out of settlement began with a survey that used a new-planned road as one of its baselines. By 1834 the road ran 150 miles from Port Credit on the shore of Lake Ontario to Tobermory on Lake Huron. The Surveyor-General, with a notable lack of imagination, called it Hurontario Street, its made-upness recalling the names of those 'great ways' that spanned England a millennium before.

To begin, the surveyors laid out a series of 200-acre squares which were then subdivided into 100-acre lots for grants and concessions (Fig. 19). The team that cast this vast grid over the landscape used a 66-foot chain to measure the sides of squares and hammered in stakes at their corners. The stakes were marked with the 'Broad Arrow' sign to indicate property of the British Crown and the Royal Cypher of King George III. 'Between every fifth lot would be a one chain or 66-foot-wide road allowance that would be staked out.'[25] Periodically the surveyor made notes about the terrain and its resources and forwarded them to the British government. The surveyors were 'delighted with the lay of the land, the quality of the timber, the richness of the soil, and its stability for farming'.[26] The survey team was preceded by an 'Advance Man', who pioneered a path for the packmen and cook and found camping places for stopovers. He recalled: 'The only wild animals we noticed while surveying was [sic] a den of young wolves, captured in a large hollow log.' The surveyors slew the cubs, later presenting their scalps in exchange for a government bounty, and passed the night listening to the old she-wolf howling for her young.[27]

Fitting individual lives into history's flows and processes can be tricky, not least because of the temptation to assume that personal stories are a result of the processes. At first sight, for instance, the departure of Mrs and Mrs Albert Carradice from Liverpool for Quebec on 29 August 1914 fits several well-known patterns. Albert was going to a new life as

19. *Extract from map of Nassagaweya township: Ontario townships were normally laid out as rectangles which were divided into numbered concessions, in turn divided into 200-acre lots which could be subdivided into smaller parcels.*

a dairyman in British Columbia – a typical destiny at that time – while family legend afterwards said that his marriage to the 21-year-old Ennis Whiteley on 4 August had been hastened by her pregnancy. That looks possible. Afterwards, no one back in Yorkshire was sure about the exact age of the Carradices' daughter, Mary; Ennis's father, William Lumb Whiteley, was a teetotal mill-owner and a trustee of the Methodist chapel at Pool-in-Wharfedale; and Ennis's sudden exit looks odd in the context of a Wharfedale-hefted family, all the rest of whose members continued to live within walking distance of each other for decades afterwards.

A closer look reveals a more intricate story. In the first place, a passenger list of the RMS *Empress of Britain* finds Albert Carradice stepping ashore at Liverpool on 16 July 1914 following a voyage *from* Canada. Love at first sight is all very well, but Albert and Ennis could hardly have met, become engaged, organised their wedding, arranged Ennis's expatriation and discovered her pregnancy all in the space of nineteen days. Evidently, Albert and Ennis already knew each other. A letter dated 1909 puts Albert at Gamsworth House, a farm in Upper Wharfedale then run by Jane Holmes, the widowed mother of William Lumb Whiteley's wife and Ennis's grandmother (Fig. 20). Ennis was then sixteen, and she, too, stayed at the farm in that year. The daughter of Ennis and Albert was born on 10 June 1910, three days before Ennis's seventeenth birthday. They called her Mary. It looks as though the birth took place at Gamsworth, although it was not registered until fifty years later. In December 1910 Albert left for Vancouver. Four months later the census of 1911 finds Ennis back with her parents at Pool, and infant Mary in the hands of Albert's mother, Bertha Carradice, in a house in Skipton. The following spring Bertha's eldest son, Francis, left for Canada, explaining his plan to join Albert in Vancouver to run a dairy farm. Four months later, Bertha took her daughter, Alice, and baby, Mary, to join them. Albert's return to marry Ennis in July 1914 thus followed an absence of three and a half years, for over three of which Ennis and her daughter had been apart.

What are we to make of these glimpses? It is tempting to bridge the gaps with speculation – for example, about the likelihood that William Lumb Whiteley imposed a blackout on the matter to maintain appearances. This would square with what we know of him from younger relatives: an authoritarian parent, paternalist mill-owner, and a Sabbatarian given to the ostentatious production of a large pocket watch in chapel if the preacher went on too long. It would also fit with a

*20. Jane Holmes, relatives and friends at Gamsworth, c. 1908. Ennis
Whiteley (seated with dog) sits in front of Elizabeth Craven (farm
manager's wife), behind whom stands Harry Cockram (see Fig. 5) flanked
by John Craven (farm manager). Jane Holmes sits beside Elizabeth Craven.
Beside her are Jane and William Whiteley. Wade Hustwick, a Bradford
journalist and local historian, stands on the right.*

continuing secrecy that denied cousins knowledge of each other's birth-
days for another hundred years. There again, when did Ennis's parents
discover that the baby had been born? Were they aware throughout, or
were Jane Holmes and her farm servant at Gamsworth able to cover
it up? When, how and by whom was the decision taken to remove the
baby from the scene, first by putting her in Bertha's hands and then by
taking her to Canada? Who paid for that journey? Was it continuing
parental opposition that caused Albert Carradice to postpone his return
to marry Ennis until a few weeks after her twenty-first birthday, when
she became free to marry from choice, or had he simply been working
hard to save up and make the marriage possible?

The truth is, we do not know, and where it is possible to check, what
was being said at the time does not always tally with what was actually
happening. Thus, Francis told Canadian immigration officials of his
plan to be a dairy farmer, whereas on arrival in Vancouver he got a

job as a steam engineer at a goldmine about 250 miles to the north. Albert, likewise, had pursued mining and gold prospecting; for a time, he worked a cinnabar deposit, digging it by hand, pulverising the ore and heating it to extract mercury. Bertha told the authorities that she was going to live with her sons at their dairy farm, which at that date did not exist. There is even a question whether in 1914 Ennis thought she was emigrating, or simply leaving for a time with a view to eventual return. Here she is seven years later writing to her mother:

> Life would be very sweet if only I could run in and see you all at a moment's notice just when I liked to ... just hope it won't be long. Albert seems to think we are better off here as in England things are heavily taxed now aren't they ...[28]

Or again: 'Things come back into my mind at times now and I feel I'd love to put my arms around your neck and say I'm sorry.'

Ennis retained her Yorkshire sense of thrift. At Gamsworth trees had been valuable, whereas Canadian timber was so plentiful that everyone used it for cooking and heating: 'you would think it a shame to burn so much good sound wood'. The year before they had bought a cow. Ennis chose her, perhaps drawing on experience gained back at Gamsworth. She is 'such a beauty', Ennis told her mother:

> She gives such lovely milk too I haven't bought any butter since we got her that will be a month she has been milking and sell 8 qts of milk a day also and feeds her calf which is a little jem [sic].[29]

Albert's nephew recalled 'quite a few cows in Vancouver' during his childhood. They 'wandered the streets and vacant lots. People kept them in their back yards.'[30] Ennis mentions Mary and Albert going out on summer evenings to cut long grass along street verges to make hay to feed the cow in winter. 'It grows very thick and soon adds up.' Ennis's delight in the cow and calf were such that she told her mother: 'I should like to ship them with us when we come back.' But they never did come back.

The dreamed-of dairy business eventually materialised but did not last. Frank's son recalled 'our basement being full of old milk bottles ... I don't know if they sold the business or went bankrupt but I got the impression my father did not have much money.' Frank loved reading and opened a book store – a notable step for one who had left school aged twelve to work as an errand boy. His son reminisced about Uncle Albert, whose lifestyle puzzled him:

He lived in a big house in the high-class end of town and drove a big car. We lived on the east side and did not have a car. During visits to his place he would show me his mining equipment … and tell stories about his mining and prospecting days … Most of the time I knew him I believe he worked for the city of Vancouver in various working level capacities on sewer and road construction as well as some gardening. He did not seem to have any high paying jobs and I always wondered how they maintained their life style … Did Ennis get some support from her family?[31]

Let us take a closer look at that family and its surroundings. Taking the Holmeses first, their farm at Gamsworth was deep Yorkshire. It is still there: a broad-built eighteenth-century house and attached barn on a grassy terrace of the River Wharfe at the foot of a fell between Bolton and Appletreewick, a village of seventeenth-century houses that is as lovely as its name. Fishermen are drawn to this reach of the Wharfe, which is fringed by shores of gleaming sand and shingle, and fed by peaty becks the colour of beer that tumble through gills and over waterfalls. The valley's flanks are full of intricacy: ravines, springs, shakeholes, sinkholes, swallowholes and nearby Skyreholme Folds, an arched back of layered limestone where mineral veins attracted miners.[32] J. M. W. Turner worked here in the summer of 1808. William Wordsworth's visit the year before inspired *The White Doe of Rylstone*, his narrative poem of the Rising of the North which begins just down the valley, where 'bells ring loud' from 'Bolton's old monastic tower'. From the twelfth century, the area formed part of the domain of Bolton Priory, an Augustinian house with mind-bogglingly extensive sheep-farming interests that reshaped Wharfedale's landscapes by the formation of new drove-ways, roads and granges, denuding the surrounding fells of woodland, and stimulating nearby markets for linen, boots and shoes. After Bolton's dissolution in 1539 its lands were sold to Henry Clifford, whose last descendant's marriage brought the estate into the hands of the Devonshire family. The Holmes family had been tenants at Gamsworth since the 1620s. Their account book survives, its prosaic lists of improvements, rents, payments and livestock intercalated with extracts from the *Iliad* neatly transcribed in Greek by a Holmes forebear who championed education for the children of labourers.[33]

Gamsworth was a microcosm of rural England before the Great War: a place where experienced tenant farmers leased their land from an estate owned by a family whose members they knew. When John Holmes died in 1878 his widow, Jane, continued the tenancy, carrying on far into her eighties, running the farm with the help of a farm manager who lived in. Jane's grandchildren came to stay in the school holidays, joining in with haymaking and muck spreading. The opening of Bolton Abbey railway station in 1888 brought visitors from nearby Bradford, some of whom lodged at the farm to escape city smoke and added to farm income. Jane Holmes's life began during the reign of George IV. When it ended early in 1914, her daughter, Jane, now Mrs William Whiteley, took over the tenancy.

The Whiteleys had arrived forty years before, in the nearby hamlet of Skyreholme – a row of houses beside a beck fed by waters that pour out of a limestone ravine called Troller's Gill. A water-powered mill was working here early in the nineteenth century. In the 1830s and '40s it was used for cotton spinning.[34] By 1850 the mill was vacant and over the next twenty years there were successive efforts to sell or let it.[35] In 1874 it was bought by the partnership of William Whiteley and Thomas Lumb. The Whiteleys were an industrious but undercapitalised paper-making family who had pursued earlier ventures at sites near Halifax and then Horsforth on the edge of Leeds; Thomas Lumb was born to Hannah Lumb before her marriage to William Whiteley in 1849. Together they converted Skyreholme mill to the manufacture of brown packing paper and millboard. It was this project that brought William's eleven-year-old son, William Lumb Whiteley, into Wharfedale, and so led to his meeting and later courtship of the girl across the fields at Gamsworth. On his walks to meet young Jane Holmes he passed through a delectable landscape of elongated grassy hillocks, between which Stangs Lane curved from side to side.

There were a lot of mouths to feed at Skyreholme. The 1881 census reveals the business supporting Thomas and his wife, Mary, their six children, William and Hannah, eighteen-year-old William, his sister, two elder brothers, Benjamin and Samuel, both of whom were married with children, and another relative. There were also tensions. In 1877 the partnership between William Whiteley and Thomas Lumb was dissolved, and although the resulting business was known as William Whiteley and Son, for practical purposes it was increasingly controlled by Thomas Lumb.[36] William Jnr and his two brothers, Benjamin and Samuel, accordingly looked to strike out on their own. In 1886 they took a lease on a mill at Pool-in-Wharfedale further down the dale that made sugar bag paper. At first this was another touch-and-go operation. However, one of the uses for their glaze paper was the wadding of .303

cartridges and rifle grenades, for which from 1914 there was large and
sustained demand. More than this, Whiteley's press-paper possessed
qualities of electrical insulation that were ideal for use in transformers
and cable windings. Advances in the generation and distribution of
electricity were then rapid, and the firm prospered accordingly. When
William and Jane married in 1887 they had lodged with a relative, then
rented a one-bedroom terrace house. By August 1919 they were able to
move into a large detached house standing in 1,350 yards of ground in a
village outside Leeds. Two months later the tenancy of Gamsworth was
surrendered and the farm's livestock and equipment were auctioned.[37]

'THREE CENTURIES IN THE SAME HOUSE,' ran a head-
line in the *Wharefedale and Airedale Observer*.[38] 'BARDEN FAMILY
IN ONE HOUSE SINCE THE EARLY STUARTS,' said the
Craven Herald.[39] William Lumb Whiteley's condition had advanced
from poverty to wealthy manufacturer in half a century. At first sight
the exchange of pastoral Gamsworth for prosperous urban fringe looks
like part of that process. However, while the giving up of the farm might
look like industry turning its back on the countryside, Whiteley's own
archive gives a more elegiac impression. He and Jane kept the inventory
of the auction and the correspondence that led up to it, with its lists of
stock detailing the months when cows were due, the calves with their
ages in months, the sixty-six Scotch ewes and forty head of poultry
and geese, the narrow-wheeled cart, the trap, the new hay chopper, the
scythes and rakes, the 400 yards 'of well-won hay to be eaten on', and
the seventy-two acres of winter eatage. They cherished the eighteenth-
century account book, and the rental agreement between Lionel Holmes
and the estate, written in a fine copperplate hand, signed on Lady Day
1845, with field names like Crabtree Garth, Lamb Close and Summer
Hill Close. They commissioned a painting, in which a young lady walks
down the slope near the barn in May or June, when swallows and swifts
skim the Wharfe for insects. The severance, said the *Craven Herald*, had
been 'almost tragic in its incidence', and that is how the record still feels.

One fact connecting most of the families in this chapter is that they
were poor, often very poor, when we first meet them. The Whiteleys
were no exception: William Lumb Whiteley's grandfather was working
at the age of eleven and received little education. His grandchildren, on
the other hand, were sent to public school. His eldest grandson, David,
joined the Royal Air Force during the second great war, and along
with about 92,000 others was sent to be trained in Canada's quiet skies
and predictable weather. By a strange symmetry, while he was among

the descendants of Yorkshire settlers in New Brunswick, and took the transcontinental train to visit Aunt Ennis during a long leave,[40] some 25,000 Canadians were settled in Yorkshire. Their badge was a maple leaf superimposed on Yorkshire's white rose, and they formed No. 6 (Royal Canadian Air Force) Bomber Group.[41] By the war's end the Canadians were spread between eleven airfields, most of them dotted up the Vales of York and Mowbray between a line from Harrogate to York and the Tees, with the remainder headquartered at Allerton Park, a Gothic revival house near Knaresborough that 6 Group's public relations officer dubbed 'Castle Dismal'. Allerton (pronounced Ollerton) was requisitioned in autumn 1942 from William Marmaduke Stourton, the 23rd Baron Mowbray, described by 6 Group's commander as 'the worst pessimist I have met for a man of forty-seven, no patriotism and full of himself and his troubles'.[42] Before handing it over, the Mowbrays held a final party, to which Lady Mowbray's brother-in-law brought two American infantry officers as guests. One of them described the scene to his parents:

> We were asked to spend the weekend at the country seat of the 23rd Lord Mowbray and Stourton. The whole thing was like a fantastic expedition into early nineteenth century England; Lord M & S has the undisputed and enviable right to kiss the King on the cheek at sight. The house is a hideous great pile of Early Victorian gloom built on the site of the original castle, which burnt down. The grounds are incredibly lovely, including a) a large herd of deer;[43] b) a curious domelike structure built by some duke to celebrate a minor victory;[44] c) a garden chock-a-block full of incredibly delicious fruit; and d) an artificial lake black with duck . . . My friend George and I were a trifle dismayed when we saw all this – we had visions of sitting glumly round the throne, speaking only when spoken to, and dying for a drink. But not at all. After a short period of iciness, while the English aristocracy were obviously wondering what species of barbarians these Americans were, a first-rate impromptu party – the best kind – started, and lasted till two in the morning.[45]

Then the RCAF moved in. Their staff included nearly a thousand women. One of Nana Wearne's daughters was among them. When on leave she could take the train to Carlin How.

Up to 2,000 people worked on each bomber station. The North Riding thus become entangled with what in effect was a series of small Canadian towns, and woven between Carlton Miniott, Ainderby

21. Members of RCAF Women's Division marching to a royal inspection in the grounds of Allerton Hall

Quernhow and Hutton Hang were names introduced by the Canadians themselves. At Croft was No. 434 Squadron, which had been adopted by the Rotary Club of Halifax in Nova Scotia (Fig. 21). Its tag was 'Blue-nose', a sobriquet for Nova Scotians. With them was No. 431 'Iroquois' Squadron. At Tholthorpe were 'Alouette' (No. 435) and 'Snowy Owl' (No. 420) squadrons. No. 429 at Leeming was 'Bison', No. 419 across the Tees was 'Moose'. And so on. Most of the airfields had been laid out in a frenzy of construction during the first years of the war, when bulldozers grubbed out hedged farmland and contractors laid triangles of concrete runways. Conditions at such places were basic. Billets, engineering and technical sites were dispersed out in surrounding countryside for safety, so that airmen spent many hours cycling back and forth between them. Much of the aircraft servicing was done outdoors, in all weathers. Men in billets awoke with bedding soaked by the condensation that dripped from the underside of corrugated iron roofs. A billet could be up to a mile from the shower block. The airfield at Skipton-on-Swale was so far off the beaten track that breweries refused to deliver to its messes, and there was no bus service to take men into York during stand-downs. Thousands of Canadians lived thus for nearly three years. While doing so they stirred the hearts of girls in quiet hamlets, thrilled their mothers with gifts of small luxuries, and boosted the takings of village pubs. Units like the Snowy Owls and the Moosemen were also communities in mourning: on average, thirty-five of No. 6 Group's aircrew died for each of the 121 weeks they were in Yorkshire. By May 1945 all they wanted was to go home.

VE Day in Yorkshire was wet. The Canadians did not care. By lunchtime celebrations were in progress in the messes at each station. Pipe bands appeared, marching around in pouring rain 'followed by all ranks in various stages of intoxication and designs of dress'. More revelry followed in the evening, often culminating in dances in the NAAFIs. Next morning was quiet as everyone slept in. The squadron commanders and their flight commanders, meanwhile, had been ordered to Castle Dismal to attend introductory talks about transatlantic flying. No. 6 Group had recently been re-equipped with new Lancaster Mk 10 aeroplanes, built at Malton Ontario, not far from the place where the wolf had howled for her lost young, clubbed by the surveyors in 1818. The plan was to fly these aircraft back to Canada and convert them for use in the war against Japan. Flying across the Atlantic in 1945 was not straightforward. For one thing, the aircraft did not have the range to fly home direct. The trip thus had to be made by stages, via Lagens in the Azores and Gander in Newfoundland. The first step was to fly to St Mawgan in western Cornwall, which was 320 miles closer to the Azores and would enable them to top up fuel. From here it was 1,200 miles to Lagens, where they would pause to refuel and the aircraft could be checked. The next leg was the longest: 1,700 miles to Gander. Then came a final stage of 650 miles to their destination, Yarmouth at the south-west end of Nova Scotia. When the crews had returned to families they had not seen for up to three years, the aircraft would be taken to Scoudouc, New Brunswick, to be prepared for their new use.

Flying over ocean called for special skills. The following days were accordingly occupied by training. Flight engineers attended lectures on engine care; wireless operators were instructed in air transport procedure; navigators practised astronavigation; pilots spent hours in the Link trainer – an early type of flight simulator. Each crew flew cross-country training flights to hone their skills. Ground crews serviced the aircraft. In slack periods they played volleyball. Throughout there was always one question: 'When do we go?'

On Saturday, 19 May, the Moosemen up at Middleton St George bade farewell to the north of England with a Victory Finale ball. This was not just any ball: it was attended by 1,200 people who danced to music played by the Streamliners, the RCAF's fifteen-piece dance orchestra, which was flown in from London. When not on tour the band's venue was the Queensbury All-Services Club, a converted theatre on Old Compton Street, north of Shaftesbury Avenue, which was used by the BBC for live broadcasts of shows and dance music. Here the Streamliners played alongside top bands like the orchestras of Jack Payne, Hal Kent, Tommy Kemp and – until his disappearance in December 1944 – the king of swing, Glenn Miller. On this evening, they played into the small hours. 'Lots of girls,' observed the squadron's war diarist. Moosemen danced with local girls, land girls, daughters of their landladies, sometimes with the landladies themselves. Also present were Canadian women officers and NCOs from the headquarters at Castle Dismal. As jitterbugging partners swung out and drew back together, the mood was anything but dismal. Towards 2 a.m. the Streamliners launched into their signature song and last dance, 'My Blue Heaven'.

> You turn to the right
> You find a little bright light
> That leads you to my blue heaven

> You find a cozy place, fireplace
> Cozy room
> A little nest that nestles where
> The roses bloom

> Just Molly and me
> And the baby make three
> Be happy in my blue heaven

Their departure was to be spread over several weeks. Aircraft would leave in small daily batches, squadron by squadron, the aim being to establish a continuous flow while avoiding congestion at intermediate airfields along the route.

The last days of May were unsettled, sometimes wet, with local thunderstorms. An advance party of ground servicing staff was flown to the Azores. Aircrew divided their kit between what could be carried as stowage and what would travel by sea. Drawers and cupboards were emptied, huts cleaned. Around the North Riding parties were organised in the Canadians' honour by locals. By the month's end kit and spares were being weighed, checked and stowed. With 3,000 miles of ocean ahead, there were dinghy drills. Crews began to pay last visits to village pubs they had come to know: the Oak Tree, the Havelock Arms, the Alice Hawthorn, the Dog and Gun, the Green Dragon.

The first aircraft flew out from Middleton St George on 31 May. The whole station turned out to watch. Waving with them were Air Chief Marshal Harris and the senior officers of No. 6 Group. Only aircrew were aboard – the much larger contingent of ground staff would return by sea on the *Queen Mary*. An airman with a cine camera stood on the roof of a van to catch the moment when the first aeroplane began to roll, with a roared cheer and waving of caps that continued as one by one they took to the air.

The Moosemen's turn came the next day. Among them was Lancaster B- Baker, captained by Flying Officer William Smith. The name 'Linden Rose' was painted on the side of the cockpit. Linden was derived from the first three letters of the names of Smith's daughter and son, Linda and Dennis, and Rose was his wife. The crew's flight engineer was Flight Sergeant Ken Munroe. Munroe's log book gives glimpses of their journey, which began with an exceptionally low pass over their favourite pub. They reached St Mawgan in Cornwall early in the afternoon and were held there for five days, 'all very cheesed', waiting for bad weather to clear. On 6 June they took off for the Azores. The flight of eight hours and thirty-five minutes was uneventful – 'navigation spot on' – and the food provided by the Americans when they got there 'was perfect'. On the 9th, they flew on to Gander, negotiated a heavy rainstorm on the way and were greeted by 'real Canadians' and a steak supper. On the 10th:

> Gander to Yarmouth. The last leg of our trip. Really poured the
> coal to old B Baker. Were met by station personnel. All crew kissed
> good old 'B'. She worked perfect all the way over. We FLEW THE
> ATLANTIC.

Linda, Dennis and Rose were there to meet them.[46]

They did not all go home, of course. Over 4,000 were dead. Among them was the 21-year-old Clarence Elsley, a member of the 'Snowy Owl' Squadron at Tholthorpe, whose surname is shared with that of a distinguished eighteenth-century Yorkshire family.[47] Elsley died on the night of 15/16 March 1944. All deaths in action are tragedies, but a peculiar melancholy attends this one.[48] He and his crew had just completed their training and had only arrived at Tholthorpe four days before. At the start of a tour it was usual for a freshman pilot to be sent on an operation as a passenger with an experienced crew to watch what happened.

On 15 March, Elsley was accordingly detailed to accompany the crew led by Flying Officer Douglas Calder for a raid on Stuttgart. They left Tholthorpe shortly after seven that evening. A night-fighter shot them down four hours later. Their aircraft crashed at Bernhausen, killing all on board. Elsley and his colleagues were buried nearby at Neuhausen. After the war their remains were transferred to the great Valhalla for airmen who had died over central and southern Germany – the war cemetery at Durnbach.

Clarence Elsley had been born in December 1920. His parents were Russell and Jeanette Elsley of Moffat, Ontario. Moffat formed part of Nassagaweya Township, another of these areas gridded out for settlement by surveyors in the nineteenth century and home to emigrants met earlier who had travelled from Danby on the North York Moors.[49] They were a small-town family. Clarence's dad ran a general store, and until Clarence entered the air force in 1942 he had worked as a clerk. His service record includes the notes made by the selection board that interviewed him. 'Poor education,' they said, and to judge from his application form, completed in rather scrawly capitals, his schooling had not taken him far. But what the assessors saw was his promise: mental ability, 'prompt reactions', 'alert, cooperative', 'will develop with training'. In answer to the question whether he might make an officer, they wrote 'possibly'. Towards the end of his training Pilot Officer Elsley was scored 80/100 for character and leadership; 'keen on his work … Should make a good captain.' His confidential personal assessment found him active-minded, cooperative and dependable. The war found qualities that might otherwise have remained latent, tantalisingly raised his prospects, and took him before they could be realised.

After the atomic bombing of Hiroshima and Nagasaki the aeroplanes brought back from Yorkshire were no longer needed. In any case, Scoudouc's sea air was doing them no good. They were ferried to the drier climate of Alberta, where they were parked on various airfields pending a decision on what to do with them. Many were eventually sold to prairie farmers, sons of Canada West immigrants, who bought them for as little as $250 and towed them away to be picked over for dynamos and jacks, or as a supply of cabling. Fuselages were adapted as hen houses; tail wheels could be fitted to threshing machines; propeller spinners turned upside down made plant pots. As the hulks settled on airless tyres, paint fading and fabric gradually disintegrating through hot summers and freezing winters,[50] the places from which they had been flown reverted to farmland.

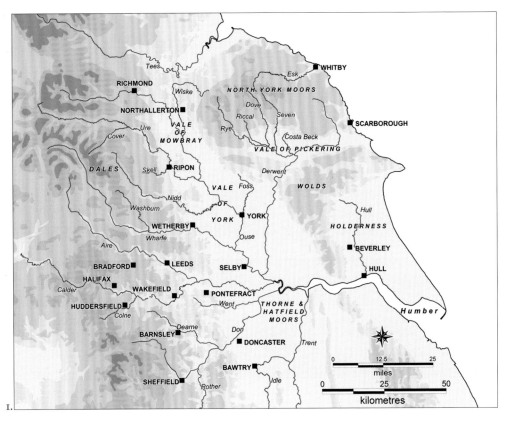

Topography of Yorkshire.

Scarborough, south bay. The site of the castle is on the promontory, upper left. William Smith's rotunda museum stands in front of the Valley Bridge, a little to the right of the V-plan Grand Hotel. The railway station with its long platforms is lower right.

3.

Richmond castle and town, by William Callow (1812–1908).

Leeds from Beeston Hill in 1816, by J. M. W. Turner (1775–1851).

4.

ABOVE: *Seventh-century jewels from grave of elite lady near Loftus.*

ABOVE RIGHT: *Eighth-century grave-marker, re-set in the north nave wall of Holy Trinity, Wensley, inscribed with the name of man called DONFR[ID]. The central square recess may have contained a semi-precious stone.*

BELOW: *Altar cloth with ammonites and snakes, church of St Hilda, Danby.*

BELOW RIGHT: *Grip and pommel of the ninth-century Gilling sword.*

9.

LEFT: *Whorlton Castle: motte and bailey remodelled in the fourteenth century with tower house, park and gardens.*

BELOW: *Pontefract Castle, surveyed in 1561. Pontefract was a pivot in defence and control of the north. It was reduced to ruin in 1649 to deprive it of strategic purpose.*

BOTTOM: *Markenfield Hall: seat of the Markenfield family, from whom it was confiscated after the Rising of the North.*

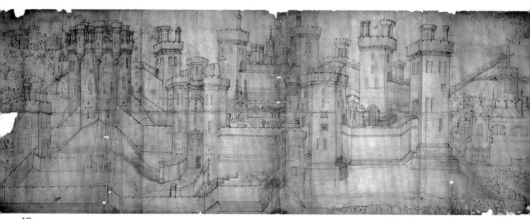

10.

11.

LEFT: *Distant view of Whitby, by Thomas Girtin (1775–1802).*

BELOW: *Bowhead whale.*

12.

BELOW: *The Dutch River with the Aire and Calder Navigation running alongside, looking north-east towards Goole and the Ouse. Inclesmoor (see Plate 16), now drained, lay to the right.*

13.

14.

BELOW: *Where the Humber begins: confluence of the Trent and Ouse flowing from left, looking west towards Goole. Part of Inclesmoor see Plate 16) lay in the angle between the two rivers.*

15.

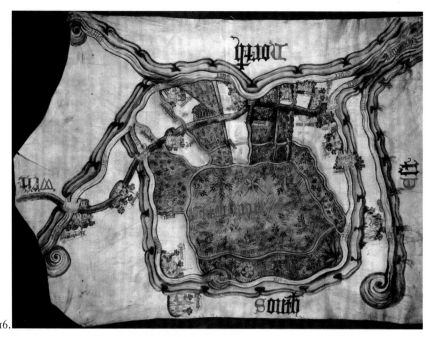

16.

ABOVE: *Inclesmoor, map of c.1450, showing a waterland bounded by the rivers Ouse, Don and Aire.*

17.

LEFT: *Beverley.*

RIGHT: *Erosion at Skipsea, on the Holderness coast.*

18.

ABOVE: *The twelfth of the last fifteen days of the world, when the stars fall from heaven: detail from Pricke of Conscience window, All Saints, North Street, York, c. 1410–20.*

RIGHT: *St George and the dragon and St Francis and the birds: Harry Stammers's 1958 window commemorating the sons of Philip and Mary Cunliffe-Lister, in church of St Mary, Masham.*

BELOW: *Cold War gaze: domes containing eastward-scanning radars at Fylingdales, c. 1964.*

20.

BELOW: *Detail from Plate 20 of birds listening to St Francis, church of St Mary, Masham.*

21.

22.

23.

LEFT: *Leeds town hall at the time of its opening, 1858. The building was designed by the Hull architect Cuthbert Broderick, who was 29 when the building was commissioned in 1852. Broderick drew on historical models in France and Italy to produce a monumental building intended to epitomise nobility, public spiritedness and civic pride.*

BELOW: *Kellingley Colliery, near Ferrybridge, in 2012. Known locally as 'Big K', Kellingley's closure in 2015 marked the end of deep coal mining in the UK.*

24.

25.

ABOVE: *Post-industrial might-have-been: visualisation of proposed* Brick Man, *Antony Gormley's 120-foot figure of re-used brick intended to stand on vacant land near Leeds railway station and be a welcome and vantage point for the city. City councillors rejected the proposal in 1988.*

RIGHT: *Bungalow, Seaside Road, Aldbrough. The building is one of several survivors of plotland chalets, most of which have been replaced by newer bungalows, along a road that has been cut by the sea. A caravan park stands opposite.*

26.

Canadian voices could often be heard in North Riding pubs in the 1970s and '80s, as veterans came back with their families to revisit old haunts or re-encounter a strange chapter in their lives when perhaps they had been someone else. The voices are heard less often these days – the youngest of the veterans are in their nineties – although some of their children and grandchildren now make the trip. Among them are the descendants of men and women who years before had set forth from some of the very places they now visit, to make new lives in places like Nassagaweya.

6

FARTHEST NORTH

And God said, Let us make man in our image, after our likeness: and let them have dominion over the fish of the sea, and over the fowl of the air, and over the cattle, and over all the earth, and over every creeping thing that creepeth upon the earth.

Genesis 1:26

O Lorde how magnified are thy works? In wysdome hast thou made them all: the earth is full of thy ryches.

So is thys great and wyde sea also, wherein are thyngs crepying innumerable, both small and greate beastes.

There go the shyppes, and there is that Leviathā[n], whom thou hast made, to take thy pastyme therin.

Psalm 104: 24–6 (translation from the Great Bible of 1539)

Calves of *Eubaelana glacialis*, the North Atlantic right whale, are about fourteen feet long at birth. The calf has been thirteen months in the womb, and immediately begins to swim. The mother–calf bond is affectionate: mothers are sometimes seen swimming on their backs, using their fins to cradle the calves on their bellies. Nursing continues for a year. Weaning begins at eight months, a little later than in human infancy, when the young whale begins to feed itself by swimming open-mouthed through swarms of zooplankton and tiny crustaceans. The whales swim slowly, averaging five knots. They are docile. For habitat, they prefer coastal waters and water bodies above continental shelves. A full-grown right whale may be fifty feet long and weigh up to seventy-nine tonnes. This is about the same as half a dozen London Routemaster buses, but unlike a bus the right whale can be frisky; it likes to breach

22. *Bowhead whale and calf drawn by William Scoresby*

and to smack its fins; it is inquisitive, given to spyhopping, and will prod floating objects out of curiosity. Females take up to nine years to reach sexual maturity; thereafter they give birth every three or four years. No one is sure how long right whales live, but lifespans of fifty years are known. This is less than the ages attainable by its larger cousin, *Balaena mysticetus*, the bowhead or Greenland right whale (Fig. 22, Plate 13).[1] The bowhead is estimated to live for over two centuries, making it possibly the longest-living mammal on the planet.[2]

The bowhead and northern right whales belong to the family of Baleen whales known as *Balaenidae*. The name derives from the plates of baleen with which they sift zooplankton from seawater. In length the bowhead varies between forty and fifty-nine feet (12–18 m) and the weight of an average adult is in the range 75–100 tonnes.[3] They are black, with a white patch on the chin and lighter colouration on the tail stock. The arch of the upper jaw rises higher than that of the North Atlantic right whale; the eyes are low-set, about a foot above the corner of the mouth. The bowhead has a pair of blowholes; when it surfaces to breathe the wide spacing of the holes gives a V-shaped blow. The blow is easily recognised at a distance.

Bowhead whales may travel alone or in social groups. Their aerial manoeuvres include breaches, tail-slapping and fin slaps.[4] Whales that come across floating logs may play with them, sometimes for long periods. Unlike northern right whales, they live entirely in Arctic waters: they seldom travel south of latitude 68°N and spend a large part of their lives above the 110-fathom depth contour, close to the edge of pack ice. Their life in far-northern waters is made possible by exceptional insulation: a bowhead carries some fifteen tonnes of blubber in a layer that averages twenty inches thick.[5]

Until the seventeenth century they wintered east and south-east of Greenland, in Baffin Bay, along the coast either side of Disko and in the Labrador Sea. In spring, they moved into the Atlantic and headed north, first to the waters around Jan Mayen Island, which was probably their calving area. The calves kept close to their mothers and were playful. Male and non-nursing females continued northwards to feeding and mating grounds west and north of Svalbard. Conditions here varied from year to year; in some years, the ice retreated early and the first whales arrived in April; in others, the pack ice broke up late, or hardly at all, and the whales would not appear in numbers until late May or early June. In autumn, they returned south. They vocalised, giving out calls that varied from a resonant humming growl to downward treble swoops like multiple violin glissandi. During spring migration they sang, sometimes at low frequencies, at others pouring out long melismatic phrases.[6]

Bowheads were not, of course, the only large creatures in the Greenland Sea. Swimming with them were *Monodon Monoceros*, the narwhal (Fig. 22), with its tusk (a projecting tooth); at the ice edge were tens of thousands of seals, walrus and polar bears that hunted them; and there was frantic life in the air as colonies of migrant auks, guillemots and kittiwakes used every second of the 24-hour daylight to raise their families before the Dark Period came back.

The seasonal cycle of light and dark governed the whereabouts and concentrations of the zooplankton upon which bowheads fed. The zooplankton in their turn were reliant on food sources such as phytoplankton, algae and ice algae, which multiply according to the proximity and thickness of sea ice.[7] Ice algae convert carbon dioxide and nutrients to organic matter and oxygen, and their release in spring by melting sea ice creates a nutritious layer at the ocean surface, available for consumption by creatures such as copepods, amphipods and mysids that form the bulk of the whales' diet. Hence, as the ice front retreated in summer, the whales moved to higher latitudes to keep abreast of the organisms at the base of the Arctic food chain.

The idea of a 'food chain' or 'food web' is one of the concepts at the base of modern ecology. It is first met in *An Account of the Arctic Regions*, a two-volume study published in 1820 by the young Yorkshire scientist and whaling captain William Scoresby (1789–1857).[8] Between 1803 and 1823 Scoresby sailed to the Greenland whale fishery in every year but one, initially as a member of his father's crew in the *Resolution*, and from 1811 (when he became twenty-one) as master. In 1813 he took command of the *Esk*, a brand-new vessel which like its smaller predecessor *Resolution* had been built at Whitby.

A typical whaling voyage began in March when vessels in the Whitby fleet (at this date, around ten) departed for the Arctic. It took about a fortnight to reach the whaling grounds. Here they would remain until August, or until the amount of whale oil they obtained reached capacity, whichever came the sooner. Scoresby used his voyages to make scientific observations as well as catch whales, and his journal entries for 16 and 17 May 1815 were of far-reaching significance.

The 16th found the *Esk* in the Greenland Sea, at latitude 77°40' north, roughly midway between the southern end of Svalbard and Greenland. The weather was fine all day, and although several whales were sighted amid patches of ice, the oar-driven whaleboats launched to chase them were unsuccessful. Scoresby used part of the day to study colours of the sea.

> The sea here is of a deep olive green colour seems thick and muddy. The appearance of the ocean is probably peculiar to Greenland. Where the sea is so deep that the rays of light appear not to penetrate to the bottom & where the water is pure or transparent, the colour is generally if not always a fine ultramarine blue.[9]

Scoresby was returning to questions he had pondered before.[10] He noted factors affecting sea colour. Among them were the texture, material and hue of the seabed, loose sand in a tideway, and outflow from a large river. He further noted that 'the sea is liable to much deception in observation', and since in places 'its colour will seem to vary with every passing cloud' it was necessary to examine the water in a way that excluded the transient influence of external light effects. He was particularly struck by areas of sea that departed from the normal clear ultramarine blue and became 'a blackish olive green' – which was its colour at the *Esk*'s current position. Moreover:

> The ice floating in this peculiar water appears yellow on its edges ... For the purpose of ascertaining its nature & submitting it to a future chemical analysis I preserved a quantity of snow from the surface of a piece of ice which had been washed by the sea and which was & is coloured to a brownish shade – this snow was still wet with sea water. On dissolving a portion in a wine glass it appeared perfectly nebulous

throughout from a number of semi-transparent spherical substances, and a vast quantity of small pieces of fine hair, like a child's hair chopped small.[11]

Scoresby put the material under a microscope. The 'globular looking substances' resolved as 'molluscous' creatures between one twentieth and a thirtieth of an inch (1.27–0.84 mm) in diameter. These were probably copepods. He found the hair-like filaments 'far more interesting'. Under high magnification they resembled necklaces, and their presence had the property of splitting light into the colours of the spectrum, which 'in some positions were very vivid and beautiful'. Scoresby had discovered krill, and as his whaler crews reboarded the *Esk* after their fruitless chase, he concluded that it was 'very evident that these substances give the peculiarity of colour to sea in these parts, and from their appearance & great profusion are evidently sufficient to occasion the great diminution of transparency which always accompanies the colour'.[12]

Next day, Scoresby returned to his investigation and made further discoveries. Using a binocular microscope he differentiated more types of plankton, and by extrapolation from the numbers present in his sample he began to estimate their quantities in the sea. In a cubic foot of seawater, he reckoned, there would be 110,592; in a cubic fathom, nearly 23 million; in a cubic mile, 23,888,000,000,000,000 – that is, 23.888 quadrillion.[13] 'What a stupendous idea does this give of the works of the creation & of the profusion of Divine Providence,' he wrote, adding that they had 'remained unnoticed for 6000 years'[14] – that is, since the beginning of the world as calculated from a literal reading of the Bible.

Scoresby realised that zooplankton formed the basis of the whales' diet,[15] and surmised that the plankton in their turn fed upon another tier of yet smaller organisms (Fig. 23). He also recognised that the fish upon which fin whales, dolphins and seals fed subsisted on zooplankton, or upon smaller fish that did so, and that the general food of the polar bear was the seal. It followed that:

the whole of the larger animals depend on these minute beings, which, until the year 1816,[16] when I first entered on the examination of the sea-water were not, I believe, known to exist in the polar seas. And thus we find a dependent chain of existence, one of the smaller links of which being destroyed, the whole must necessarily perish.[17]

This is the first recognition of a 'chain' of being on record. The terms food chain and food web entered general use only from the late 1920s.[18]

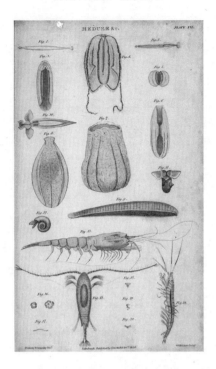

23. *Some Scoresby drawings of marine creatures observed in Arctic seawater*

Scoresby anticipated this by more than a century. When Charles Darwin boarded HMS *Beagle* eleven years later, Scoresby's *Account of the Arctic Regions* was one of the volumes in the ship's library. Darwin's own work in the coming years would contribute to a revision of the world's age from the biblically derived figure of sixty centuries to 4.54 billion years.

Sailing homewards from her 1815 Arctic voyage the *Esk* was making about five knots as she passed the Firth of Forth on the morning of Tuesday, 1 August. Around noon they overtook a smack on its way from the Orkneys to London with a cargo of fish. The smack's crew gave them the news of the Battle of Waterloo, and of Napoleon's subsequent abdication. After twelve years, Europe was at peace. Scoresby's crew responded with three heartfelt cheers. Their enthusiasm was partly patriotic, but also from a sense of relief. During the Napoleonic wars, homebound whalers had often been accosted by naval warships and crew members seized for service. A whaleman impressed into a naval crew might not see his home again for years, if ever. By the end of the eighteenth century 'the press' had become so frequent that some masters stopped short of their home ports to put most of their crew members ashore, enabling them to cover the last few miles overland.

The next day the wind fell in the afternoon, but picked up again in the evening. Scoresby marvelled at phosphorescence in the *Esk*'s wake, and 'liquid fire' where her bows cut the water. Next morning they were off the mouth of the Tees, 'the Yorkshire land within sight' and 'above 60 coasters around us'. 'At ten got sight of Whitby Abbey bearing SbE [south by east].' By late afternoon they were off Sandsend and were met by a pilot cobble (Plate 12). The pilot crew brought news of wives, children and friends. 'For my own part,' wrote Scoresby, 'I had particular cause for thankfulness in hearing of the health of my affectionate wife and son as well as parents and other kindred.' Scoresby was also told that the *Esk* was the first of the Whitby whaling fleet to return. On the morning of the 4th he went ashore by cobble, returning at noon. 'At 1 Pm. Got under weigh [sic] and immediately that the signal was displayed, indicating a sufficient water for the ship we bore her away for the harbour under a strong press of sail.' Bringing a sailing vessel into Whitby called for the exercise of simultaneous skills:

> The great velocity of the ship impelled her, head to wind, a considerable way up the harbour, the sails were speedily withdrawn, warps taken out and vigorously applied to windlass and capstern [sic], whereby we were enabled to heave through the bridge and immediately grounded. Unbent sails.

After nineteen weeks and running over 20,000 miles the *Esk* was home.

Next day the *Esk* was hauled into her berth and moored. The following Monday, Scoresby paid the seamen's wages, with components for monthly wages, cutting and striking money, and a down payment on oil money, pending boiling of the *Esk*'s cargo of blubber in one of the boiling houses along the River Esk. When this stinking process was completed and gauged, the result was 135 tons of oil.

Meanwhile, other vessels had been returning. The *Aimwell* entered harbour on the 17th, with blubber from eleven whales (Scoresby estimated this would yield 140 tons of oil), followed by the *William and Ann* (four whales), the *Volunteer* (two) and the *Resolution*, which had returned 'clean' (none). Scoresby had last seen the *Aimwell* with the *Valiant* when they had been 'pinched' in ice on 22 June. He had feared for their safety, as also for the *John*, out of Greenock, the *Henrietta* and other ships which had 'got beset at about the same time'. 'It appears that they were frequently endangered and that some of the ships received severe presses from the ice.' Scoresby was pleased to hear of the safe return of the *John*, but many of the others 'were ill rewarded for the anxiety of themselves

24. *The* Esk *in crisis, 1816*

and their friends having succeeded in capturing a very few small fish'. The *Henrietta* came home on 21 August with about 140 tons of oil. Last back was the *Lively*, with blubber from three whales together with part of the cargo salvaged from the *Clapham*, which had blown up on 15 May when an on-board fire detonated the ship's magazine. Whalers carried gunpowder for gun signalling and in case of entanglement with ice.

The besetment from which the *Aimwell* and the others had extricated themselves was trivial in comparison with the crisis faced by the *Esk* in the following year. On 29 June 1816 Scoresby discovered that part of the *Esk*'s keel had been torn off by ice. During the first forty-four hours after the accident the *Esk*'s crew pumped out seawater at a rate later calculated by Scoresby as sixty-six gallons or two tons a minute.[19] Attempts in following days to fother the ship, to unload her and make running repairs on the ice, or invert her to give carpenters access to her keel, met with no success (Fig. 24). However, her impressive stability and remaining buoyancy convinced Scoresby to strike a bargain with the master of the *John* to take her in tow for the thousand miles to Shetland in return for almost half the *Esk*'s catch. The *Esk* was barely manageable, and two days were needed merely to draw her clear of the ice. A fortnight later the two vessels arrived off Shetland, now also in the company of the *Phoenix*, a whaler from Whitby which shepherded the *Esk* for the remainder of a journey that was remembered by Scoresby as 'at once hazardous, disastrous, and interesting'.[20]

Expectancy hung over Whitby during the final weeks of waiting for the ships' return; wives, sweethearts, children and friends climbed the cliffs each evening to scan the horizon. Elizabeth Gaskell caught something of this near the opening of *Sylvia's Lovers* (1863). When Molly and Sylvia round a bend in the road that reveals the red-tiled roofs of the town below, and white-sailed fishing boats on a sapphire sea, also there, lying to by the harbour entrance, is a larger vessel. Sylvia is new to the neighbourhood:

> but Molly, as soon as her eye caught the build of it, cried out aloud
> –She's a whaler! She's a whaler home from t'Greenland seas. T' first
> this season! God bless her!' and she turned round and shook both
> Sylvia's hands in the fullness of her excitement.[21]

Gaskell wrote *Sylvia's Lovers* when Whitby's whaling days were done, and used Scoresby's *Account of the Arctic Regions* to help visualise what they had been like. During the heyday of the whale fishery, from the 1750s to 1820s, virtually everyone in the town had had some connection with the industry. Like the Arctic food chain, there was an interdependency between those who built and crewed the whalers, the suppliers of rope, cordage and canvas, dealers in casks, tools and navigational instruments, clothiers, insurance brokers, farmers and providers of provisions, and general retailers who benefited from the purchasing power of everyone else. Even trades like brickmaking and house-building prospered, witnessed in lavish villas that were put up for successful ship-owners and captains. The rhythms of later Georgian Whitby's society followed the whaling year. Winter was the time when money was spent, when there were concerts, dances and visits by travelling actors while the ships were being overhauled. In late winter the ships were readied and stocked, setting forth once more in late February or March.

In 1806 Scoresby and his father found a 'remarkable opening of the ice' north of Svalbard and used it to take the *Resolution* beyond the latitude that was normally accessible, reaching the record northing of 81° 30'. This was only 590 miles from the North Pole, and for some years Scoresby toyed with the idea of an expedition to reach it from Svalbard, travelling across the ice with sledges that could also function as boats.[22] On his return, he registered as an undergraduate at the University of Edinburgh, where for several years he combined study of science in winter with whaling in spring and summer.

The choice of Edinburgh was interesting. In part, no doubt, it was because of the city's reputation for progressive scientific inquiry – reflected in the influence of figures like James Hutton (1726–97), whose geological theory had begun to challenge biblical chronology, or Robert Jameson (1774–1854), who had taken up the university's chair of natural history two years before. But it may also have been reinforced by the fact that Whitby had a stronger sense of companionship with places like Edinburgh, Newcastle and London than it did with York or the West Riding. The wide moorland plateau which rose at its back could not be crossed at speed.[23] The dissection of this area by steep-sided valleys made it hard going for coaches, and when it was covered by snow (as during Georgian winters it often was) the direct route to Yorkshire's main towns and cities was blocked for weeks at a time. In contrast, Newcastle could be reached by sea in less than a day, and Edinburgh in two. Crew members from east-coast ports were regularly in each other's towns, making Whitby's mental neighbourhood a narrow strip, perhaps a day's walk in width and several hundred miles long.

Scoresby's teachers at Edinburgh included Jameson, the physician and chemist Thomas Hope (1766–1844), and the mathematician John Playfair (1748–1819) whose presentation of Hutton's geological theories did much to widen awareness of what they were.[24] Together they fostered the teenager's enthusiasms and gave methodological guidance that helped him to put them on a practical footing.

Scoresby's observations, sampling, record-keeping and analysis soon came to embrace all aspects of Arctic science. At the age of seventeen he was investigating optical effects at high latitude of refraction, mirages and coronas, recording their appearance in sketches and watercolours (having taken drawing lessons to help him to do so) and attempting to model their causes by calculation. Seven years of measurement of the salinity of seawater, using hydrometers with his own calibration, led him to the conclusion that Arctic seas were less saline than those in temperate latitudes. His records of sea temperature in the water column and bathymetric survey led him to discuss and model ocean currents. He was a talented and astute observer and recorder of plants, animals and animal behaviour. His meteorological observations were important both for the development of the science and for Arctic navigation. He addressed the formation of clouds, and his detailed study of the properties of ice crystals, snow, hoar frost and rime stood for years without parallel (Fig. 25). (At the age of eighteen, he had enthralled fellow crew members by carving a lens from a lump of ice and using it to focus sunlight to light their pipes.) His research into terrestrial magnetism, and the effects of local attraction when a compass was moved to different

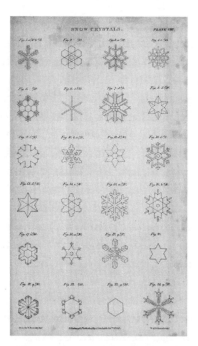

25. Extract from Scoreby's classification of snowflakes

parts of a ship, was presented to the Royal Society in 1819. In the winter of 1821/22 he began to investigate what we would now describe as the effects of electro-magnetic induction, and his paper reporting further experiments on magnetism was read to the Royal Society by Humphrey Davey two years later.[25] Other scientists of the day with whom he formed working friendships and correspondences included Michael Faraday (1791–1867), Darwin's friend Joseph Hooker (1817–1911), and the biogeographer Alexander von Humboldt (1768–1859). When the British Association for the Advancement of Science held its inaugural meeting in York in September 1831, Scoresby was present as a founder member. With him were pioneers in other fields who by some strange synchrony were near neighbours of his childhood village of Cropton. William Smith ('author of the geological map of England') of Hackness was one of them. The aeronautical engineer Sir George Cayley from Brompton was another.[26] By this time, however, Scoresby's career had taken an entirely new path.

Scoresby's sketches and triangulated surveys of Arctic coastlines culminated in a substantial achievement in 1822, when, sailing out of Liverpool in the *Baffin* (a vessel he had helped to design), he combined whaling with the survey of a large part of the east coast of Greenland.[27]

His journal entries for the final days of earlier voyages suggest a quiet delight, almost a kind of ritual, in the recognition and recording of familiar landmarks that foreshadowed home: St Abbs Head, the Staples and Farne lights, the first sight of the Tynmouth light, the precious moment when they saw 'the Yorkshire land'. Best of all came the moment when news was passed aboard about 'my beloved wife and lovely boy'.[28] But at the end of the 1822 voyage the news was otherwise: Mary had died while they were away. Scoresby made one more voyage, then quit sea-going and entered Queens' College, Cambridge, to study for the Anglican ministry. This may not have come as a complete surprise to former crew members, who were used to Sunday musters on deck to hear his evangelical sermons. Following ordination and pastoral duties in Yorkshire, Liverpool and Exeter, in 1838 the Revd William Scoresby was invited to accept the incumbency of the large, fractious, dissenting, insanitary parish of Bradford – which is where we met him near the beginning of this book.

Scoresby's decision to abandon whaling may have been strengthened by a realisation that for commercial purposes its Arctic days were numbered. During a voyage he usually kept a daily tally of the number of other vessels in sight. In the Greenland Sea, there were often three or four sail to be seen at any one time, and sometimes upwards of thirty – which was about the size of the entire combined whaling fleets that had been annually fitted out by chartered companies from the Netherlands and England back in the sixteenth century.[29] Since then, other nations had joined in. By the later seventeenth century, the annual average of whalers in Greenland–Svalbard waters was around 300; in some years, they took more than 2,000 whales; by the mid-eighteenth century the northern right whale population had collapsed. In 1817, the year of Scoresby's last voyage in the *Esk*, 150 whalers had set forth from Britain alone. Between 1810 and 1818, 824 ships sailed for the Arctic whale fisheries from England, and 361 from Scotland. In just four of those years the English vessels took 3,348 whales, besides narwhals, bears, seahorses and numberless seals.[30]

The status of the remaining population of bowhead whales was affected by climate as well as hunting. Ice cores show a Little Ice Age between the mid-sixteenth century and early twentieth, when the mean summer temperature in Svalbard was at least 1°C below the summer mean during the half-century that followed. Within this period there were alternating warmer and colder interludes. During warm spells the ice front retreated and whales could flee into leads and openings amid fragmenting pack ice. In cold periods the summer ice edge came south, and whales were penned against it, to the advantage of those who hunted them. The second half of the eighteenth century saw a rise in the number of years with closed pack ice. A lucrative period for the whalers followed. In the long run, however, the combination of many 'south ice years' and intensified hunting was lethal.[31] Northern whaling declined during the 1820s and '30s simply because it was no longer profitable. When the last Whitby whaler returned in 1837,[32] the Svalbard population of bowhead whales had been annihilated.[33] Whitby's contribution to its disappearance since 1753 had been the capture of 2,761 whales – just 2.2 per cent of the total estimated to have been killed.[34]

What had been done with them?

The bowheads' oil had been used to light streets and houses, as a lubricant for machines and ever-more numerous atmospheric engines, as an ingredient of soap, varnish and paint, and to assist the preparation of leather.[35] Their long, curving jawbones had been turned into the ribs of sheds, used as gateposts or garden pergolas, or powdered for fertiliser. The strips of baleen could turn up almost anywhere. If baleen is put in hot water it can be worked into almost any shape, and in an age before plastic its elasticity and flexibility lent a versatility that suited scores of uses: for the springing of sofas and chairs; for making fishing rods or frames for umbrellas and parasols; springs in the suspension of carriages; for corsetry, or the finer fibres to stuff mattresses.

The one function the baleen no longer performed was the one it had evolved to accomplish: the annual extraction of an estimated 3.5 million tonnes of zooplankton from northern seas. The surplus became available to other populations, of fish and birds. It was William Scoresby who taught about food chains; the equilibrium of the North Atlantic ecosystem had been utterly changed.[36]

Other Arctic products, like narwhal tusks brought back for whittling into walking sticks or hat-stands, or the myriad pieces of scrollwork and engraving undertaken by on-board scrimshanders, seem almost incidental in comparison with the killing by European and American mariners of 122,000 whales. On that basis, the bears are no more than a footnote. Polar bears prey mainly on seals, and hence are usually found near the ice front. They usually kept their distance. Opportunities to shoot them for their pelts were correspondingly limited: over eighty years, Whitby crews brought back just fifty-five skins. When a mother was shot, an attempt might be made to capture a cub, for sale to a fair or circus. If the cub survived incarceration in a barred box or barrel for the rest of the voyage, that was its future.

Scoresby, observant as always, made notes on the behaviour of a polar bear cub in the presence of its dead parent – its whimpering, nudging of the lifeless mother as if trying to wake it up, the refusal to leave its side. An observer today might find such a scene pitiful; Scoresby's notes show no such sign, no empathy. His declaration of faith at the start of his account of how whales were caught tells us why.

> The Providence of GOD is manifested in the tameness and timidity of many of the largest inhabitants of the earth and sea, whereby they fall victims to the prowess of man, and are rendered subservient to his convenience in life. And this was the design of the lower animals in their creation ... Hence, while we admire the cool and determined intrepidity of those who successfully encounter the huge mysticetus [i.e. the bowhead whale], if we are led to reflect on the source of the power by which the strength of men is rendered effectual for the mighty undertaking; our reflections must lead us to the 'Great first Cause', as the only source from whence such power could be derived.[37]

In a literal reading of Genesis, whales and all other creatures were subservient to human purposes by the will of God. Hence, far from entertaining any sense of reservation about killing them, it was God's gift of strength to man and a gentle temperament to whales that enabled the men, usually, to kill them. This was a different kind of food chain. It was also just as well, for capturing a bowhead whale was a mighty undertaking, which occasionally began by the killing of a calf to attract its mother. Just once, Scoresby confessed: 'There is something extremely painful about the destruction of a whale, when thus evincing a degree of affectionate regard for its offspring, that would do honour to the superior intelligence of human beings.' But he got over it: 'the object of the adventure, the value of the prize, the joy of capture, cannot be sacrificed to feelings of compassion'.[38]

When a whale was sighted, whale boats were launched from the mother ship. Each was crewed by rowers, a harpooner, a line manager and a steerer (who used an oar rather than a fixed rudder, to minimise drag). A bowhead would normally spend no more than two minutes at the surface to breathe, whereafter it would dive and remain underwater for up to twenty minutes. It was impossible to predict where it would resurface, and while there could be clues, like eddies in the water or the behaviour of seabirds watching overhead, at this stage it was a matter of

chance whether contact could be sustained. The ideal was to approach
the whale from the rear and drive a harpoon into its body, behind the
head. The effective range for a hand-thrown harpoon was up to ten
yards, or about thirty if the harpoon was fired from a gun.

The purpose of harpooning the whale was not to kill it but to attach
the boat to the animal and retain contact through what followed. The
wounded whale now made convulsive efforts to escape, threshing with
its twenty-foot flukes – which could overturn or damage the whale-
boat – and then diving. The whale now became a 'fast fish', submerging,
swimming either downwards or just under the surface, or seeking shelter
under sea ice if any lay nearby. On occasion a whale would swim under
ice until the whole of the lines of one boat had been run out. The boat
in turn now became a 'fast boat', and if ice of great width intervened the
whale could escape. A technique to retard the flight of the whale was
to cast two or three turns of line around a post fixed close to the boat's
stem, holding this close so that friction slowed it down. In the process
the post could become so hot that the wood smouldered.

Other boats gathered; as the whale weakened the strategy was to close
in and drive in more harpoons. Now began the actual killing: a frenzy
in which crew members repeatedly jabbed and stabbed with lances. The
surrounding sea was dyed red. The whale blew a mixture of blood, air
and mucus from its two blow holes. Imminent death was signalled when
the blow turned to pure blood and, as the whalemen said, 'the chimney
went afire'.[39]

Usually, there was a final convulsion; the tail reared and swirled; at the
end, the whale turned on its back or side.

All this could take anything from thirty minutes to half a day.[40] It is
just possible that there is a bowhead alive today that witnessed it.[41]

The rest – the three cheers, the slow tow back to the ship, hacking a
notch round the whale's girth to fasten it in a sling, the blubber spades,
the specksioneer clambering about with his crampons, the emptying of
water ballast from casks to make way for cubes of diced blubber (Fig.
26), the feasting by fulmars and sharks after the release of the creature's
ruins – let us leave the rest for another time.[42]

Whitby today is one of Yorkshire's favourite places. Atkinson Grimshaw's paintings of the harbour after dark, water reflecting the glow from
shop windows, sails and masts silhouetted, recall it in the generation
after Scoresby. The harbour is still the centre of the red-roofed town,
and it still works, though these days its craft are mainly for leisure and
fishing, and most of the fish are caught within a few miles. Alongside

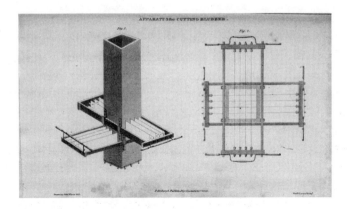

26. *Blubber slicer, 1820*

the sea-food cafés, the fossil and jet shops, the Goths in dark eyeliner, there are still signs of whaling. At the top of West Cliff is a whalebone arch.[43] If you climb the 199 steps and enter the churchyard of St Mary you can find memorials to seamen who never came home. From here you also get a good view of the museum at Pannett Park, across the Esk, where blubber spades, harpoons, flinching knives, and saws for cutting different kinds of ice are on display.

Whitby remembers the whalemen's skills – their tenacity and courage; the expertise of navigators; physical and financial risk-taking, fortunes made and lost; Homeric journeys; the cohesion conferred upon an isolated community by an industry's amalgam of anticipation and uncertainty. The place to catch the spirit of that is on the abbey headland, where from late July in Scoresby's day Whitby people took evening walks to scan the sea up past Kettleness. Here, you can search the distance for a home-bound sail, or think about the silence that is the requiem canticle of a bowhead's song.

EAST

The East Riding lies mostly north of the Humber and west of the River Derwent. It is predominantly open. Low-lying farmland claimed from marsh and bog is found beside the Humber's headwaters. East of this rise rounded, undulating chalk hills dissected by dry valleys. These are the Wolds, steep-sided to the north and west, and tilted gently eastward until they disappear beneath the low-lying plain of Holderness.

The Wold valleys were formed maybe twelve thousand years ago when water spilling from melting ice carved through frozen chalk and precipitated the collapse of solution hollows. Now and then some of them run again, as if reawakened, dampened by winterbournes, flushes and gypsies. Widely spaced villages bear names like Fridaythorpe and Warter. Between them are expanses of arable so vast that unimproved grassland is now mostly restricted to roadside verges and churchyards, where if you are lucky you may see Marbled White butterflies roosting on grass stems or fluttering amid knapweed and thistles in July.

East Riding places have distinct personalities. If you are in Hull, even in its interwar suburbs, you could never imagine yourself to be anywhere else. The minster of red-roofed Beverley is one of the most calming places you can enter. On winter afternoons, jackdaws caw around the steeples of churches rebuilt in the nineteenth century by owners of great estates. Patches of creased ground mark the sites of villages that were abandoned six centuries ago. Beyond it all is Holderness, where meres, carrs and ghosts of marshes are dotted through rich farmland.

Today's East Riding of Yorkshire was formed in 1996. Its area is not far from that of the historic county, minus Hull and some parishes around Goole, and part of the Vale of York west of Pocklington.

7

HUMBER

'I'm not safe in a room with maps,' Winifred Holtby wrote to a friend in 1926. 'They go to my head like Vermuth – the only forms of intoxicant that really give me pleasure. Vermuth *and* maps, I mean. I like both.'[1]

When Holtby started work on *South Riding* in 1934 she drew a map of her imagined county, which was the East Riding with selected places and features renamed (Fig. 27).[2] The Humber and its headwater, the Ouse, became the River Leame. She sketched the Leame's channel westwards up to the point where it is crossed by a railway. This corresponds with

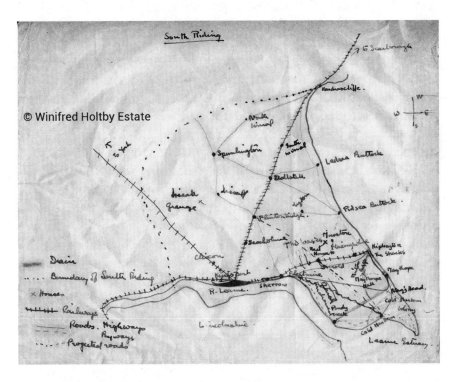

27. Winifred Holtby's sketch map of South Riding

the bridge at Goole, whence the line from Doncaster runs past Salt-marshe station to join the main line for Hull at Gilberdyke – a candidate for the 'harsh-named halt' that Philip Larkin had in mind when he was on his way to the 'surprise of a large town' in the poem 'Here'.[3]

By some esoteric coincidence, the place where Holtby's map leaves off corresponds with the point where another remarkable representation begins (Plate 16). This is a sumptuous coloured map of Marshland, Yorkshire's south-east corner, that was painted on a calfskin around 1450.[4] The map was made to illustrate a legal dispute and covers an area bounded by the rivers Ouse, Aire, Don and Trent. It shows villages, timber-framed houses, waterways, stone-built bridges, causeways, way-side crosses and ancient woodland, and at its centre is the depiction of a delectably vegetated fenland, with greens, hints of gold and luxuriant plant tendrils akin to some William Morris textile. The medieval map-maker named this area Inclesmoor. The little left of it today is known as Thorne Moors.[5]

Inclesmoor originally covered about 240 square miles, and it is a sur-prise that so many of them were in Yorkshire. They include land along the south bank of the Ouse as far as its confluence with the Trent, which instinct tells you ought to be in Lincolnshire; however, when you arrive at the Trent's west bank you are still in Yorkshire. In this out-of-the-way corner is a village called Adlingfleet, Æthelyngflet, 'the prince's water-way', which was the easternmost place in the West Riding until the reorganisation of local government in 1974. Today, after nineteen years' exile in unlamented Humberside, it is in East Yorkshire. Just beyond, the waters of the Ouse and Trent flow together, and the tidal estuary we call the Humber begins (Plate 15).

The Humber drains a fifth of all England. Places as far away as Bir-mingham, Stoke-on-Trent and Leicester are within its catchment. In the early Middle Ages it bounded kingdoms and it still forms part of the divide between the ecclesiastical provinces of Canterbury and York.[6] Reed beds and grazing marsh near Adlingfleet make a quiet place to reflect on the extent of its influence. The reeds are home to avocet and bittern, quartered by marsh harriers looking for frogs and small mam-mals, and only just downstream the estuary has already widened to a mile. To reach the Humber's Yorkshire shore we first need to go back, close to the point where Inclesmoor and South Riding touch, to the road bridge at Booth.

Booth Bridge was opened in 1929. Before that there was a ferry (Fig. 28). Scheduled ferry services connected Hull and Lincolnshire from

1820, while the arrival of Ermine Street on Humber's south bank about fourteen miles downstream at Wintringham (next to a place that was called Ferriby by 1086) points to the existence of a ferry link to the north bank town of Brough at least from *c.* AD 70.[7] Chunks of Roman masonry recycled in the base of Wintringham's church tower imply the former presence of some nearby lofty structure. This might have been a beacon, but the stones belonged to a cornice that was so large that it must have been intended to be seen from far below. When the Ninth Legion arrived here in the third quarter of the first century, did it build a monumental arch to mark the edge of empire?

But back to Booth, which for ordinary travellers was the first place west of Hull where the tidal Ouse was safely and practically crossable by oar-driven boats, and the lowest crossing point available to travellers from the north.[8] Booth's ferry was thus well used. People, carts, stage coaches, livestock, even cars were manoeuvred onto broad low-slung craft propelled by watermen who sculled over the bow or stern (Fig. 28). To cross by road required a ten-mile detour further inland to Selby. The 1929 bridge at Booth was a swing-bridge, allowing vessels to pass upriver to Selby. Selby is over fifty miles from the North Sea, and it is almost as much of a surprise to find that it was an international port as to discover that pods of porpoises used to swim through it in pursuit of salmon. Selby's maritime heyday followed the opening in 1778 of a canal that linked the town with Leeds and Wakefield, enabling the transhipment of cargoes to vessels that could go straight to sea (Fig. 29).

There are more inland ports hereabouts, the result of successive adjustments and extensions to the waterways that extended like capillaries into the West and East Ridings (Figs. 30, 31). Docks at Goole, just downstream from Booth, opened in 1827 following the development of a new canal link from Knottingley. Goole became a destination for serpentine trains of coal-carrying box barges known as Tom Puddings (Fig. 32). On arrival each barge would be lifted by a hydraulic boat hoist and its contents tipped into a coaster. The opening of a canal connecting the Don Navigation with the Trent aided Thorne's status as a place where manufactured goods from Rotherham and Sheffield could be transferred to seagoing vessels that then threaded their way out of Yorkshire and away to east-coast ports or the continent. In parallel, far-from-the-sea Thorne became a shipbuilding town that produced sloops, schooners and latterly seagoing steamboats and tugs.

After the Napoleonic wars Yorkshire's waterways bustled with craft.

28. *Booth Ferry*

29. Selby, seen in 1932, is fifty miles from the sea. Ships were built here from the late nineteenth century until 1992. Other inland shipyards existed at Thorne (Richard Dunston) and Grovehill (Cook, Welton and Gemmell) near Beverley on the River Hull: see Fig. 33.

30. Decorative ironwork on the twelfth-century door of St Helen, Stillingfleet, includes a representation of Adam and Eve (top left) and Noah's Ark, visualised as a contemporary longship (originally with mast and rigging, reconstructed from positions of nails). This kind of medieval vessel plied the tidal Ouse to nearby York. It was the ancestor of the Humber keel – a type of blunt-bowed manoeuverable craft with a square sail that came to work Yorkshire's canalised rivers and waterways (see Figs. 31, 49).

31. Clinker-built keel on the Driffield navigation. Many keels were built to suit the lock size of particular waterways.

32. Tom Pudding train on Aire and Calder Navigation

On one day in 1828, for instance, the *York Herald* recorded the arrival in Goole of fifty-eight coastal vessels (from London, Sheerness, Wisbech and Boston), and forty-four departed, thirty-four of them for ports around the Wash, the remainder to Newcastle, Yarmouth, Maldon and Kent. On the same day, a foreign-owned vessel sailed for Hamburg, two others left for ports in Scandinavia, and Selby saw fourteen arrivals and thirty-seven sailings. There were also steam packets that left Hull daily for Gainsborough, Goole, Thorne and Selby.[9] Names of the vessels – like *Two Brothers, Nine Brothers, Charlotte and Eliza, William and Mary, Providence, Amity, Hopewell, Success, Active, Humility, Hercules, Adonis* – evoke late-Hanoverian preoccupations with family, loyalty, good luck attending virtue, and classical mythology.

While thinking of names, across the Ouse on Thorne Moors, the bog pimpernel, shore sedge, great sundew, the scaly male fern 'Cristata', marsh St John's wort, bog asphodel, milk parsley, lesser wintergreen, white beak sedge, and pod grass were just then dwindling towards local extinction. When it was mapped in the fifteenth century Thorne was part of a wetland complex of mire, fen and marsh around the head of the Humber that extended for hundreds of square miles.[10] At its heart were several raised mires – domes of wet peat irrigated by rainfall – and a mere described by John Leland as being 'full of good fish and fowl' and as 'almost a mile over, a mile or more'.[11] Marshland supported a plenitude of birds, insects and plants, and in 1626 Charles I commissioned the Dutch engineer Cornelius Vermuyden to drain it.

Vermuyden responded by diverting the flow of the southern branch of the River Don into its northern channel. The southern Don had hitherto wandered across marshland in a series of lazy loops to a rendezvous with the Trent just south of Adlingfleet. The northern course into which all the Don's waters were now directed had already been straightened. The diversion left the southern Don to fade, and led to flooding in the north. Vermuyden dealt with the flooding by cutting a new channel, the Dutch River, that enters the Ouse at Old Goole (Plate 14).[12] Over the next three centuries further episodes of drainage, changes in land use practice, and the advent of steam-powered pumps converted marshland into farmland.[13] The mere is long gone, and twentieth-century suburban enthusiasm for gardening accelerated the mining of peat from the raised mires. Hence the requiem for bog pimpernel and great sundew. It could have been worse. In 1972 Dr Edmund Marshall, Member of Parliament for Goole, spoke in favour of an international airport on Thorne, citing a recent editorial in the *Goole Times* that had described it as 'a bald patch of nothing in a featureless countryside – a stretch of sour, waterlogged land'.[14]

Thorne and Hatfield Moors are the only surviving mires of their kind in Britain, yet the merest morsel of what once there was. Their peacocks, cranes and spoonbills have gone, the Old Don is no more than an occasional damp-stained ghost that wanders through windy unhedged fields, and the bitterns that were once so commonplace as to earn a local name ('butter bump' – say it very fast and it sounds a bit like the bittern's call) are now seldom seen here outside nature reserves. Even so, the moors are still home to 5,000 invertebrate species. Maiden's blush, the barred chestnut and round winged muslin are three out of 674 species of moth that have been recorded, and the ruddy darter, four-spotted chaser, and southern hawker are among the dragonflies that skim over abandoned peat workings. Cranberry, bog rosemary and round-leaved sundew have begun to recolonise.[15] Nightjars nest. You may hear a nightingale or see ruff. In winter, a gliding hen harrier may scan the ground. But the marshland that the Inclesmoor map-maker knew has gone. Its disappearance is often called 'reclamation', as if Vermuyden and his successors were restoring order to an area that had hitherto been pointless. Heading east, past Goole's still-busy docks, pausing to admire a Tom Pudding Hoist, the lofty listed hydraulic accumulator tower, and the noble goods office of the sometime Lancashire and Yorkshire Railway, the sight of stands of turbines looming above the Ouse levee invites the thought that 'reclamation' here is a euphemism for ecological tragedy.[16]

Heading east takes you through hamlets behind the river's floodbank. Saltmarshe, Cotness, Yokefleet, Blacktoft, Faxfleet and Broomfleet are quiet, brick-built places which co-exist with occasional Georgian mansions, remnant windmill towers, bungalows with fake leadlights and barren gardens, and granges amid trees planted for shelter at the ends of private tracks. The floodbank hides the river. At Faxfleet the new-born, mile-wide Humber might as well not be there, unless you climb the bank and discover its brown 'level drifting breadth'.[17] Faxfleet and Broomfleet look out across Whitten Sand towards Lincolnshire. *Whitensandes* was one of many treacherous shoals between the lower Ouse and Humber entrance that were described by the chronicler Roger of Howden (fl. 1174–1201) in his sailing directions for those going on crusade from York to Jerusalem.[18]

Until the 1770s the district aback of these places was another now-lost wetland: a 5,000-acre marshy common called Wallingfen, once laced by wriggly tidal creeks that fed the River Foulness, since desiccated by reclaiming improvers and overwritten by miles of hedgeless rectangular fields, straight-ruled roads, and farm drive entrances framed by solitary brick pillars with incongruous carriage lamps.[19] Remnants of logboats used by prehistoric traders turn up in places where creeks once fished by butter bumps are today blank farmland.[20]

Water is the constant. Just inland from Brough, the next place east, successor to the Roman ferry station where the Humber begins to cut through the chalk, is the Roman villa of Brantingham. In the fourth century, the villa's owner commissioned a mosaic pavement with a goddess of good fortune at its centre. Around her reclined eight water nymphs.[21] Why water? Water, like trees, or the way birds flew, 'provided contact with the divine or demonic'.[22] Why eight? Did they represent the Humber's tributaries and headwaters – the Ancholme, Trent, Don, Aire, Ouse, Derwent, Foulness and Hull?

On indeed to Hull, the city that surprised Larkin with its 'domes and statues, spires and cranes', where silence was 'laid like a carpet' on a Friday night in the Royal Station Hotel and there were 'ships up streets'.[23] Prince's Dock, Humber Dock and originally Queen's Dock extended inside the town. As H. V. Morton observed:

Ships sail right into the heart of Hull. They saunter casually across the main streets, their masts become mixed up with the electric cable poles. Trawlers steam in from the North Sea across roads and nestle their smoke stacks against the chimneys of Hull. Barges roll in casually, with the skipper smoking his pipe and looking up pleasantly at the long line of taxicabs, oil-cake wagons, cement carts, and tram-cars which wait respectfully for the bridge to swing back. [24]

One of Larkin's colleagues recalled that when he arrived in the 1950s you could find your way around the city by its smells. One family of wafts drifted from the varnish, paint and colour works along the River Hull. A 'reasty stench' emanated from the oil-extracting and cattlecake mills at Stoneferry. Yet more stinks stemmed from Homco Oil, the Soap Works, Pearson's Disinfectant Works, Reckitt's Starch, the tannery, brewery, seed-crushing mill, and from different reeks of 'the fishing', such as wood-smoking, cod liver oil and fish meal.[25] There was also an aleatory soundworld composed from horn blasts of Humber shipping, sirens, hooters, steam whistles that signalled the start and finish of industrial shifts, and bicycle bells. Hull's level terrain made it ideal for cycling.

> ... the main noise was the ringing of bicycle bells as workers pedalled off in the morning, at dinnertime and teatime, to and from their homes and queued up six deep at the dozen or so level crossings which cut across the town's roads, until the trains had whooshed past and the gates opened ... At night, if you lived in the fishing district, there would be the clatter of clogs as bobbers and filleters went to their work on the fish dock, but all over the town you could hear the baleful call of the fog-horn for, although Hull was cleaner than many industrial towns, most people had coal fires and the factory chimneys belched out endless clouds of smoke.[26]

But not now. The Spurn lightship is moored today in the marina as an exhibit, and the *Arctic Corsair*, the city's last sidewinder trawler, floats quietly in the River Hull, explored by visitors on guided tours. The *Corsair* was built upriver at Beverley in 1960; she was one of the 'ships up streets'. The visitors wonder at her mighty Mirrlees diesel, her adventures in the cod wars (when the *Corsair* rammed an Icelandic fishery protection vessel), life on deck under ice-crusted rigging bringing in a catch, the narrowness of her companionways, the bakelite switches, wardrobe-like crew spaces and high wooden parapets to restrain sailors from being thrown out of their bunks during high seas. Above all they

33. Launch of the trawler Vera *at the inland shipyard at Grovehill, near Beverley, 1907*

may wonder at her size, for the *Corsair* is surprisingly hefty – bigger, say, than the *Cutty Sark*, most of her forward part being a hold for the fish.

Photographs of pre-war Hull show densely packed long-roofed warehouses and workshops running back from the waterfront into a town honeycombed by narrow streets and yards (Fig. 34). The roofscape is redolent of continental cities like Rostock and Lübeck. And there we have it, Hull as part of an alliance of trading guilds in north European coastal cities: the Hansa. We shall meet this again in the mercantile world of the patrons of the windows of All Saints, North Street (Chapter 9). During the reign of Elizabeth I, Baltic iron, flax, fish, tar and salt were among the commodities carried up the Humber and thence up the Ouse in boats that returned with lead from the Dales. Many a new-built Jacobean country house was panelled with Baltic boards, while knives and edged tools manufactured in Hallamshire passed through inland ports like Bawtry and Doncaster to Hull, for shipping to the east.[27]

Daniel Defoe caught these near and far horizons when he visited Hull in about 1720. He thought the town smaller than Hamburg, Danzig or Rotterdam, but probably busier than 'any town of its bigness in Europe'. And this was partly because it stood downstream of so many confluent rivers flowing out of the north and Midlands:

all the trade at Leeds, Wakefield and Halifax . . . is transacted here, and the goods are ship'd here by merchants of Hull; all the lead trade

34. Queen's Dock, Guildhall and neighbourhood, Hull, 1931

of Derbyshire and Nottinghamshire, from Bautry Wharf, the butter of the East and North Riding, brought down the Ouse to York: The cheese brought down the Trent from Stafford, Warwick and Cheshire, and the corn from all the counties adjacent, are brought down and shipp'd off here.

In Defoe's opinion, Hull's status was also a reflection of the 'fairer character' of its merchants and the 'justice of their dealings':

They drive a great trade here to Norway, and to the Baltic, and an important trade to Danzig, Riga, Narva and Petersburg; from whence they make large returns in iron, copper, hemp, flax, canvas, pot-ashes, Muscovy linen and yarn ... all which they get vent for in the country to an exceeding quantity. They have also a great important of wine, linen, oil, fruit, etc. trading to Holland, France and Spain ...

Other east-coast towns with Hanseatic warehouses included Boston, Lynn and Yarmouth. The destinations of all those sloops, brigs and schooners that plied to and fro from Goole and Selby in the early nineteenth century were continuations of earlier patterns of trade.

Shipping news around the middle of the nineteenth century gives a flavour of the city's ever-widening horizons. Under 'Foreign Arrived' and 'Foreign Sailed' on one day in 1858, for instance, the ports from which ships had set forth or for which they departed were not just neighbours like Rotterdam, Antwerp or Hamburg, but such places as Riga, Odessa, St Petersburg, Sulina in Romania, Quebec and Havana.[28] Small wonder, then, that by the 1880s Hull had become Yorkshire's most cosmopolitan city, with communities from Russia, Scandinavia, Italy and Poland, or that there are places called Hull in at least twelve of the United States.[29] People who live in other parts of England often speak of Hull as somewhere remote, on the edge of things, out-of-the-way. This says more about England than about Hull.

> On and out, then, into Holderness, where
> Wind-muscled wheatfields wash round villages,
> Their churches half-submerged in leaf. They lie
> Drowned in high summer, cartways and cottages,
> The soft huge blaze of ash-blue sea close by.
>
> Snow-thickened winter days are yet more still:
> Farms fold in fields, their single lamps come on,
> Tall church-towers parley, airily audible,
> Howden and Beverley, Hedon and Patrington.[30]

35. Harvest at Skidby, Yorkshire Wolds, 1947

36. Sleeping guards below niche for Easter Sepulchre, church of St Patrick, Patrington

Beyond Hull, Patrington's parish church is the Queen of Holderness, who watches over Yorkshire's furthest corner. Pevsner thought her one of the most beautiful churches in England.[31] Her grace is found not only in the presence of her spire, but also in the life-enhancing observation of inner detail, like the three soldiers carved in relief below the platform where the sacred elements would be placed on Good Friday to await their resurrection. The men guard Christ's tomb. They are mail-clad, helmeted and armed as soldiers in the early stages of the Hundred Years War (Fig. 36). And they are asleep, lolling in their niches, one with his head propped by an upraised arm. Six centuries later, Aircraftman 2nd Class Edward Hughes sat nearby, chin cupped in hands, reading Jung and Shakespeare, counting down the days to his release from National Service.[32]

Patrington's snoring guards were contemporaries of the last days of the nearby port of Ravenser Odd – '*Hrafn's* gravel bank[33] – a strange place which had been founded on a Humber island by the Count of Aumale around 1235. When Edward I awarded it a borough charter in 1299 it was already a fair-sized town,[34] while its position at the Humber entrance enabled it to forestall traffic that would otherwise have gone upriver to Hedon or Hull. Luckily for them, Ravenser was drowned by storms and disappeared in the 1360s.

Holderness is rich in upward-reaching structures. Many of them have to do with the sea, like Withernsea lighthouse – tall, slim, white and tapering – that rises out of the town, or the Humber-facing lighthouse at Paull, built in 1836 by Trinity House, with its wrought-iron balcony and adjoining keepers' houses.[35] Ever-shifting sandbanks reduced Paull's utility and prompted the addition of two lights at Thorngumbald Clough, set in quirky rocket-like towers perched on metal stanchions. And with quirkiness in mind, what of Admiral Storr's tower, built on a knoll in 1750 to give the retired admiral a view of passing vessels;[36] or Bettison's folly, a cylindrical tower made of 'treacle bricks' at Hornsea;[37] or the 1920s water tower at Mappleton? But of all the towers, none are so evocative as those on Spurn Point, the apex of Hull's 'three-cornered hinterland',[38] where a succession of lighthouses has gleamed since the fifteenth century.[39]

Spurn exists in a kind of dynamic equilibrium – unstable, an island at some times and peninsula at others, changing shape by the tidal hour, its relationship with Holderness transforming,[40] migrating westwards by about eight inches each year, public yet secluded. Ambiguity extends to its formation, about which there are differing theories,[41] while historically its fluctuating communities make a kind of mirror to different preoccupations at different times – like liminal monasticism,[42] pilotage and rescue,[43] telegraphing of ship movements to distant audiences, wartime defence, nature conservation,[44] or neighbour to Holtby's Cold Harbour, where ex-servicemen toiled to earn a living off the land.

Spurn is not a uniform promontory. There is the Narrow Neck between Kilnsea and the head; Chalk Bank; the Middle Bents near the root of the Neck; the Greedy Gut – a shallow muddy channel on the Humber side, or the Stony Binks, a shoal that branches off the point at a tangent. For centuries, the Binks were used for gravelling – the excavation by hand of gravel which was shovelled into gravel sloops that rested on the bank when the tide was low and taken away when the tide returned. Spurn gravel was used for turnpike roads across the East Riding, the construction of Immingham docks, and as ballast for ocean-going vessels that sailed out of the Humber. Thousands of tons of Spurn gravel lie on the floor of the Greenland Sea, where they were jettisoned by Hull whalers. Another Spurn commodity that travelled far and wide was the cobblestone. Fist-sized stones left by ice were extracted for urban roadmaking and as building material for houses, barns and churches, this use being much furthered by the taxation of bricks between the later eighteenth and mid-nineteenth century.[45]

Spurn cobbles were laid on end and tilted, giving local walls a distinctive herringbone texture

Spurn's nearest neighbour is the hamlet of Kilnsea, which was relocated in the nineteenth century when its predecessor fell into the sea. The sea hereabouts gnaws away yards of Yorkshire each year (Plate 18). In the mid-seventeenth century Kilnsea's medieval church of St Helen was said to have been 'well situate'. A century later it was measured ninety-five yards from the cliff edge. By 1824 the cliff had arrived at the chancel, which collapsed, followed by part of the steeple in 1828 and the rest in 1831.[46] As the sea washed out the churchyard, the bones of former parishioners tumbled onto the beach. At Owthorne, another churchyard taken by the sea, a visitor found robins nesting inside a skull; the parents popped in and out through an orbit peering emptily out of the boulder clay to feed their nestlings. Like Hull, its face was 'half turned to Europe',[47] its back to the Humber, 'king of all the floods'.[48]

Being half turned towards something is also half looking away. Hull, with its Hanseatic heart, oceanic outlook and optimistic interwar suburban planning remains one of Britain's most singular cities. It has also become one of the most deprived. In the later eighteenth and nineteenth centuries, Hull, next to London, was England's main point of immigration from the continent. Among those who arrived and settled were Danes, Dutch, and Jews escaping discriminatory taxation in Prussia, violence in Ukraine, and (from 1772) Russia's incorporation of much of Poland.[49] On 24 June 2016 two-thirds of Hull's people voted to leave the European Union.

8

WONDERFUL AMY

The aeroplane ran lightly over the turf, drawing dark wheelmarks along the sheen of dew. Then it danced, brushing the daisies, cleared the low hawthorn-sprinkled hedge, and was away up into the clear sweet air.[1]

Thus does Sarah Burton take to the air at the start of the Epilogue of *South Riding*. It is half-past six on the morning of 6 May 1935, the day of George V's Silver Jubilee. Her fellow passengers are a photographer from the *Kingsport Chronicle* and a journalist who is gathering material about Jubilee decorations across the Riding. The pilot sets course first for Hardrascliffe – the Bridlington of Holtby's imagined county.

If there was a place for this scene in Winifred Holtby's mind's eye it was very likely Salt End, on the outskirts of Hedon. Salt End was the site of Hull's municipal aerodrome, a roughly rectangular seventy-eight-acre grass field bordered by a low hawthorn hedge, and when Holtby began work on the book in the spring of 1934 she passed it every week on the train that ran from Withernsea to Hull.[2] On airshow days the train stopped at nearby Hedon Halt, an unstaffed platform left over from a time when the field had been used as a racecourse. Other sites seen from the line appear in *South Riding*. One of them is the White Hall at Winestead, a small country house near Patrington that became the model for Robert Carne's hall at Maythorpe.

Hull's aerodrome was established in 1929.[3] Two years later Councillor Frederick Till, chairman of the Hull Development Committee, spoke of the city's 'great advantages from the point of view of flying'. Councillor Till might have stepped straight out of, or indeed into, *South Riding*, which Holtby described as 'a sort of comedy of local government'.[4] Till believed that Hull 'possessed a magnificent aerodrome that was going to go ahead in the world of aviation'. Air services to the continent

37. Hedon aerodrome, 1931

were in prospect; the council wanted to make Hull 'a real flying city'.[5]

An aerial photograph taken in 1931 tells another story (Fig. 37). There is not much to see – just a bungalow-like clubhouse, three garage-style fuel pumps, a hut and a couple of aircraft sheds. Two light aircraft stand in front of the larger shed. A closer look reveals wheel tracks in the grass, and the outlines and ridges of older fields. A car on the otherwise empty Staithes Lane is stopped at the level crossing. The nearby signal is down for a train on its way to Withernsea.

The photograph was one of several taken a few minutes apart. Another, from very low level, gives a closer view of the parked aircraft, about which more later; the train to Withernsea shuffles steam as it passes beyond; passengers in the rear carriage peer out at the photographer's aircraft skimming towards them (Fig. 38). A third image shows a hay-wain bearing mown grass by the airfield entrance,[6] the mock-Elizabethan exterior of the Wireless Receiving Office,[7] and the tower of All Saints, Preston, on the horizon (Fig. 39). This is more Deep England than Le Corbusier's exhilarant world of flight wherein 'everything is scrapped in a year'.

A time there was when Hull had been Britain's third seaport. In the

38. Hedon aerodrome a few minutes later: a Civilian Coupe (L) and de Havilland Puss Moth (R) are parked before the shed; a train bound for Withernsea passes beyond

late 1920s an airport seemed necessary to carry that tradition into the future. The port journal argued for swift action to steal a march on rivals like Manchester and Liverpool.[8] City fathers did their utmost to boost commercial interest in their field. Profile-raising events were organised, like the air pageant in April 1930 at which Sir Sefton Brancker, the UK's Director Civil Aviation, proclaimed a 'tremendous market' for civil flying.[9] In January 1934 Alderman Benno Pearlman, former Lord Mayor, chairman of Hull's Aerodrome Committee and another candidate for *South Riding*, spoke of his ambition for Hedon to become 'the Croydon of the north'.[10] A few weeks before, it had been reported:

> Further information about the proposed K.L.M. Hull–Amsterdam air service [...] is to hand. Consideration is being given to a possible midday service from Hull in addition to the 7 a.m. departure. This would enable morning services from Manchester, Newcastle, Liverpool, etc., to connect so that passengers might reach Copenhagen or Malmo the same day. It is hoped that the fast connection between Hull and the Continent will make it worthwhile for air-transport

39. Hedon aerodrome from the south-west, 1931

concerns in Britain to operate connecting lines from various big business centres, such as Glasgow, Belfast, Dublin, Manchester, Liverpool, Birmingham and Newcastle.[11]

At the end of May 1934 KLM did indeed instate a daily Hull–Amsterdam service, with which several feeder services were meant to connect.[12] Alderman and Mrs Pearlman and KLM's general manager were among the passengers on that first flight, which was met by city dignitaries and a headline in the *Hull Daily Mail*.[13] But at this date the vagaries of weather, primitive equipment and underdeveloped operating methods meant that few civil air services were sustainable without subsidy. By December the KLM service had been suspended. It was restored for a time in 1935, and there was optimism that the newly installed wireless service would be available to Provincial Airways, which opened a route to 'England's Riviera' (aka Torquay),[14] and Aberdeen Airways, which was planning a service from London to north-east Scotland via Hull and Newcastle.[15] However, the Torquay service failed, as soon afterwards did Provincial Airways, and the east-coast plan remained in abeyance.

The only service that seems to have approached practicality in this period was an air ferry across the Humber estuary. For a gannet down from Bempton, Hull and Grimsby are just sixteen miles apart. For everyone else the trip was laborious. Until the opening of the Humber Bridge in 1981 the distance by road was eighty miles, beginning with a drive in the opposite direction to Goole and then south-eastward to cross the Trent. The Humber ferry and train took one and a half hours. In contrast an aeroplane did it in fifteen minutes and even with the taxi ride from Hull it was still under half an hour. Following experiment,[16] the service opened in July 1933, initially with three flights each way per day.[17] By September demand was sufficient to make the service hourly. Once again, however, the underlying economics were unfavourable and after a time the air ferry was discontinued. By now, moreover, the city's Airport Committee had come to realise that its airport was too small for impending developments in civil aviation, and that the main road and railway on its edges made it impossible to enlarge.[18]

Hull Municipal Airport closed at the end of the 1930s. Few remember it, and those who do often do so because of events on one day: Monday, 11 August 1930, when Hull-born Amy Johnson returned in triumph following her solo flight to Australia. Film cameramen were at Hedon when she arrived overhead in *Jason*, her green and silver DH 60 Moth at 4 p.m. Jason was the registered trade mark of her father's company. The footage shows what followed.[19] Waiting for her that windy Monday were her father – John 'Bill' Johnson, a fat cigarette hanging from his lips – uniformed footmen bearing bouquets, Councillor Richard Richardson the Lord Mayor, the city sheriff, and thousands who had come out from the city and Holderness. Johnson taxied *Jason* before the thirty-deep waving crowd, brought the aircraft to a stand amid photographers and their ladder-bearers, and switched off.

Johnson hoisted herself onto the cockpit rim and waved. Nimbly avoiding the ardent arms of Councillor Richardson she jumped down amid worthies and heavily be-hatted wives. Here began a transformation: when Sir Arthur Atkinson, president of the Hull Aero Club, invited her onto a platform from which they could speak (amplified by 'Marconiphone'), the crowd could see that beneath the leather flying coat Amy wore high heels and a dress cut just below the knee. Hull was 'tense with excitement'. The six miles to the city centre were decorated with bunting and lined with a third of a million people, who pressed forward to shout greetings as the motorcade passed.[20] As the cars approached the city centre, the crowds thickened until all traffic 'was at a standstill and every window was filled with sightseers'.[21]

Next day the Hull Rotary Club (of which Amy's father was a

prominent member) gave a formal lunch. The menu (including iced melon, chicken fricassee, cold 'bouffet', rice pudding and trifles) was overprinted on a stylised map of Johnson's route to Australia. The occasion included a performance of a 'Rotary Anthem to Rotarian J. W. Johnson's distinguished Daughter'.

Amy Johnson had reached Darwin on Saturday, 24 May. The lyricist J. G. Gilbert and composer Lawrence Wright exploited the euphoria that followed by penning a popular song:

> Amy, wonderful Amy,
> How can you blame me for loving you?
> Since you've won the praise of every nation
> You have filled my heart with admiration
> Amy, wonderful Amy,
> I'm proud of the way you flew,
> Believe me, Amy, you cannot blame me, Amy
> For falling in love with you.

Gilbert and Wright worked fast – Jack Hylton and his orchestra recorded the song only nine days after Johnson's arrival.[22] However, this was not what was sung at the Rotary lunch. The Rotarians had written their own anthem and enlisted Mrs Bertram Wadsworth Shaw to sing its first verse:

> How d'ye do, Amy Johnson, how d'ye do?
> We extend a heart welcome now to you;
>   We would like to be allowed
>   To say that Rotary is proud
> And your praises sing aloud. How d'ye do?

The Rotarians then joined in, in harmony:

> How d'ye do, Amy Johnson, how d'ye do?
> To Britain's great traditions you were true;
>   None your feat will 'ere surpass,
>   It was in the highest class,
> You're a champion Yorkshire lass. How d'ye do?

The rhyming of 'class' with 'lass' suggests that the Rotarians were 'very Yorkshire'.

> How d'ye do, Plucky Johnnie, how d'ye do?
> You achieved the task you set yourself to do;

To the job in hand you stuck,
Though you had a bit of luck
It was mainly Yorkshire pluck that pulled you through.[23]

Later in the day Johnson addressed a Rally of Youth in the City Hall, telling 3,000 girls and boys to 'abandon safety first', cultivate self-reliance and enable England to lead the world in flight as in centuries past it had led by sea. 'Our country, England, must come before safety.'[24]

Johnson's message of national strength through technical progress, youth and fearlessness conformed to the policy of the *Daily Mail*, which had been encouraging airmindedness through prizes and sponsorship since before the Great War (Fig. 40).[25] The newspaper had co-sponsored Johnson's voyage, kept the public abreast of her daily progress under headlines like 'No News for 24 Hours' and 'Sea Flight Peril', and contracted her to write about flying and tour the country to put herself and *Jason* before the public.[26] The values they symbolised had been blazoned six days before at a *Mail*-organised Savoy luncheon in Johnson's honour. Guests were selected to celebrate success across British arts, sport, science and culture. Among them were the Olympic athlete Harold Abrahams,[27] the engineer and television inventor John Logie Baird, composer and actor Ivor Novello, conductor Malcolm Sargent, authors J. B. Priestley and Evelyn Waugh, and the aircraft manufacturer

40. *Henri Salmet – one of the* 'Daily Mail *airmen' – in a Blériot XI at South Bay, Scarborough, 1914. Salmet gave exhibition flights at Hedon racecourse – the site of Hull's municipal-airport-to-be.*

Sir Alliott Verdon Roe. Esmond Harmsworth, 2nd Viscount Rothermere, press magnate and former Conservative politician, presided at the top table, where Amy Johnson's neighbours included Sir Sefton Brancker, who had been responsible for Lord Wakefield's support for her flight.[28] Also there were R. C. Sherriff, whose play *Journey's End* was then running at the Savoy Theatre; Noel Coward, on the eve of the premiere of *Private Lives*; and Marjorie Foster, the markswoman who had just out-competed nearly 1,000 (mostly male) entrants to win the King's Prize for rifle-shooting. A greater muster of national talent at this time is hard to imagine, and like a fairy at the top of a Christmas tree the *Mail* placed Johnson at its apex.

Johnson's air tour of the kingdom began on 26 August.[29] It was meant to last two months and visit thirty places. Her speeches proclaimed the need for England to lead the world in aviation. At this distance it is hard to tell if the message was hers or Rothermere's, but either way it does not seem very profound. A closer look, however, finds it to be part of something larger.

The call to Hull's young to abandon caution had echoed Rothermere, who had said in his speech at the Savoy that '"Safety first" is not the emblem inscribed on Miss Amy Johnson's shield, nor I hope upon the shield of England.' He contextualised this in two following statements: first, England had a destiny, in which it was necessary to have faith, and second, obstacles to attainment of that destiny would be overcome not by logic or system but by 'character, faith and courage'.[30] The idea of world leadership in aviation was thus invested with a kind of moral quality, and that in turn drew upon a perception of aviators as an exclusive fellowship possessed of outstanding abilities and intensified faculties.[31] Gender, aeroplanes and risk thus coalesced as an embodiment of national purpose.

It is a question how far Johnson went along with this. Back in May, on arrival in Burma, she had said: 'This is just an ordinary flight, except that it is longer. Every woman will be doing this in five years' time.'[32] Similarly, while it would be natural to assume that her support for the United Empire Party at the Paddington South by-election in October 1930 was prompted by allegiance to Rothermere, earlier correspondence with her father shows that the original impulse was the other way about.[33] Rothermere's UEP stood for free trade within the British Empire; back in February Johnson realised that her projected flight would connect different parts of the Empire, and identified the UEP as a potential sponsor.

Johnson's tour was abandoned ten days after it started. The *Mail* announced that for medical reasons she had been ordered to 'take a prolonged rest'.[34] Rothermere had introduced her to England's elite as someone 'born and brought up in the City of Hull ... unknown, without much money, without much influence',[35] which was true, but downplayed the underlying position. Her father was partner in a successful and growing business,[36] her mother, a music teacher, was from a prominent city family. When Johnson set her sights on flying to Australia her parents contributed £800 to buy the aeroplane. Johnson had been well educated, took a degree in Economics, Latin and French at the University of Sheffield, and was athletic. She was also versatile. One of the main reasons she reached Australia was her understanding of how her aeroplane was built and why it flew. During 1929 she worked at the de Havilland factory airfield at Stag Lane, north London, learning to take down and erect engines, maintain fuel systems and make repairs to airframes. Unafraid of oily hands and torn nails, when things went wrong on her voyage, she fixed them (Fig. 41).

Johnson's victorious return to Hull on 11 August coincided with another event that connected Yorkshire and civil aviation. At the end of July the newly built R100 airship had set forth on a proving flight

41. Amy Johnson at Stag Lane, Edgware

from her home station at Cardington in Bedfordshire to Montreal. On 11 August she toured Ontario's skies, arriving over Toronto during the morning rush hour and bringing the city to a standstill. She went on to Niagara Falls, then back to Montreal. Two days later R100 departed, returning to Cardington in under fifty-seven hours. The airship had been built in a vast shed left over from the First World War on the flat lands north of Howden, in Yorkshire. Her visit to Canada caused a sensation. In distance the two countries were far apart; in time, their separation was now down to hours.

Elation soon turned to sorrow. R100 was one of a pair of government-sponsored experimental airships that had been developed during the later 1920s with the aim of bettering communication across the British Empire. R100 had been built by a subsidiary of Vickers. The other, R101, was designed by the state's Royal Airship Works at Cardington. R101 underwent modification between June and 1 October to tackle a weight problem. Flight trials of the amended vessel were curtailed to meet a political deadline: the six-week Imperial Conference opened on 1 October, and, since airship development was on its agenda, Lord Thomson, the Secretary of State for Air, was keen to report on R101's performance before the conference ended. Late in the afternoon of Saturday, 4 October, R101 set forth with fifty-four passengers and crew on a proving flight to India. Early next morning she crashed and burned on a hillside near Beauvais in northern France. Just eight of those aboard survived, two of whom died soon afterwards. Among the dead were Lord Thomson, and Amy Johnson's encourager and proselytiser for municipal aerodromes, Sefton Brancker.

In Yorkshire the calamity had immediate consequences. R100 was mothballed pending a public inquiry, and then scrapped when the government abandoned airships in the wake of the inquiry's report.[37] This resulted in the immediate lay-off of the staff who had built her at Howden. Among them were the gifted designer Hessell Tiltman (1891–1975) and the engineer and novelist Nevil Shute Norway (1899–1960). Their response to redundancy was to start a new company, Airspeed, which found premises in a former municipal bus garage in York.

Airspeed's story is entwined with Johnson's, and with the ways of local government. The Depression was still in progress yet, despite the promise of new employment, negotiations with the county borough of York over the bus garage 'seemed interminable'.[38] An even larger challenge was to attract capital from potential investors. Tiltman and Norway wanted to be pioneering. This was partly because this was

their instinct, but also because established companies enjoyed economies of scale, and trying to emulate the kinds of machine they built
would leave Airspeed at a commercial disadvantage. In investors' eyes,
however, a design philosophy founded on originality spells risk, and the
response from potential shareholders was meagre. Amy Johnson was
one of those who did invest, as did the Yorkshire banker Ralph Beckett,
Lord Grimthorpe, who became chairman of the board. Grimthorpe
combined banking with breeding racehorses and the sport of bobsleigh.
Airspeed was all of a piece with such chancy pastimes: on a number of
occasions during the company's first years Grimthorpe's guarantee was
all that stood between it and bankruptcy.

Two designs emerged from Airspeed at York before the company
outgrew its garage and relocated to Portsmouth. One of them was a
machine built at the prompting of Sir Alan Cobham, a tireless proselytiser for flying and a founder member of Airspeed's board. Cobham
was planning a series of National Aviation Days in which a gaggle of
different aircraft would call at places across the country to introduce the
public to flight through displays of aerobatics, gliding, parachute des-
cents and introductory passenger trips. For the last he was looking for
two or three aircraft capable of carrying groups of around ten people;
such machines would need to be simple to maintain and withstand an
outdoor existence in all weathers. Airspeed's answer was the Ferry: a
three-engine biplane christened *Youth of Britain* which joined Cobham's
fleet in 1932 and had carried nearly 92,000 people by the year's end.[39]
The Ferry was preceded by the Tern, a high-performance sailplane
which was begun to raise profile while the company waited to move
into the garage. The Tern took to the air in August 1931, and in the
hands of Carli Magersuppe gained the British distance record a few
days later. Magersuppe was young but very experienced, and already
accustomed to flying from a launch site on Stoup Brow, a steep hillside
overlooking the cliffs of Ravenscar and Robin Hood's Bay. From here
he covered the eight miles to Scarborough, where a thousand-strong
crowd watched his arrival overhead and landing on the beach of
North Bay.[40]

Airspeed's first and in some ways most inventive early idea never
progressed beyond the concept stage: it was for a light aeroplane with
the interior of a car. Most aircraft hitherto had been open and noisy,
and called for protective clothing for those who flew them. Tiltman
visualised a machine with an enclosed cockpit, comfortable seats and
sliding roof. A closer look made them think again: the de Havilland

Moth family now dominated the market for private flying. And, in any case, such an aeroplane was already being built – in Hedon.

The photograph of August 1931 with the Withernsea train passing in the background shows what it looked like (Fig. 38). The nearer of the two parked aeroplanes has an enclosed cockpit, with all-round glazing, side doors, a draught-free cockpit and luggage compartment. It was called the Coupé and cost £650 – rather less than a large upmarket family saloon. Its builders were the Civilian Aircraft Company, which from January 1931 occupied a long single-storey brick building on the southern edge of Hedon aerodrome. An advertisement described the Coupé as 'trimmed like a car', with 'sociable seating' and innovations like wheel brakes; it was the 'first enclosed two-seater monoplane in England'. The machine could be flown by someone in shirt sleeves or a city suit. Only a handful was built: de Havilland was working along similar lines, and an example of the result stands beside the Coupé in the photograph: a Puss Moth. As Tiltman and Norway foresaw, the capacity of a larger, established company was likely to make for irresistible competition.

The moving spirit behind Civilian was a Yorkshireman called Harold Boultbee (1886–1967). Boultbee was born in Eston, where Benjamin Myers had brought his family from Bradford a few years before, and where Harold's father was the vicar. At the end of the nineteenth century the family lived in Marske-by-the-Sea, a village between Redcar and Saltburn, and it may have been here that his interest in aeronautical design was sparked or reinforced. At any rate, in May 1910 Robert Blackburn, an aviation pioneer from Leeds,[41] used the broad sands between Saltburn and Marske to try out a monoplane of his design. By the following year Blackburn had established a flying school on the beach at Saltburn.

Boultbee advanced through the aircraft industry, successively working for Bristol, the Gloucestershire Aircraft Company and Handley Page.[42] In the early 1920s he was for a time immersed in a project that philosophically prefigured the Coupé: an enclosed motor scooter that was so advanced in engineering and finish that it could be mistaken for a vehicle of the 1970s. It was called the Unibus and was claimed to run without noise or vibration. The idea of 'the car on two wheels', as it was marketed, was that a businessman or travelling professional like a doctor could ride it without being splashed or muddied. The Unibus's only reported drawback was a price of 95 guineas, around £3,870 at present values. Like the Coupé, it was bold and forward-looking but did not find its path.

A week after the reception for Amy Johnson in Hull a representa-
tive of the Civilian company visited Hedon to assess the building on
Hedon aerodrome that might become their factory.[43] Also that week, a
Mr John Kealey of Ribblesdale Road in Streatham wrote to the *Daily
Mail* asking if any of its readers had noticed that it was a Yorkshire
girl (Amy Johnson) who was the first woman flyer to Australia, and
a Yorkshire man (James Cook) who discovered Australia. Mr Kealey
concluded: 'Good for Yorkshire.'[44] He could have added that discovery
of the laws that enabled Johnson to fly could be traced to the work of
another Yorkshireman, George Cayley, in the days of George III.

Cayley was born in Scarborough a few days before the end of 1773.
His father was a baronet, a member of the lowest stratum of England's
hereditary aristocracy. The family seat was High Hall, a small country
house outside the village of Brompton-by-Sawdon, between Scarborough and Pickering, and Cayley succeeded to it as the 6th baronet at
the age of nineteen. If he has a place in general memory today it is as
a slightly eccentric figure who in 1853 launched his coachman across
Brompton Dale in a home-made glider. In fact his interest in aero-
dynamics was well advanced by the early 1800s, and was well supported
by the mathematical education he had received at the hands of several
accomplished private tutors. By 1804, the year before Trafalgar, Cayley's
attainments already included the measurement of wing lift, and a viable
glider – the first aeroplane in the world.[45]

Cayley combined practical experiment with theoretical analysis.
Technical data in his notebooks and papers on aerial navigation enable
us to trace the evolving relationship between the two. He identified
the now-familiar principles that attend the flight of a rigid aircraft: lift,
weight, thrust and drag – or, as he put it, how 'to make a surface support
a given weight by the application of power to the resistance of air'.[46]
As early as 1801 he was seeking to give mathematical expression to the
relationship between lift and the angle of wing incidence. He studied
questions of stability, control and propulsion. Previous experimenters
had usually been empiricists who assumed the need to combine propulsion and lift in flapping wings, like birds; Cayley saw the need to
treat the two separately. He built a number of gliders (1804, 1818, 1849)
and in 1843 published a conspectus of his earlier aeronautical achieve-
ments. Some of his contentions have since been shown to be mistaken;
nonetheless, he arrived at 'many of the basic aerodynamic, structural and
stability features of the successful aeroplane, tested them, and showed
that they worked'.[47]

One of Cayley's daughters reported her father's view that 'a day passed without acquiring a new idea was a day wasted'.[48] Cayley, like his correspondent Charles Babbage, father of the programmable mechanical calculator, was a polymath whose inquiries and commitments ran in many directions. At various times, for instance, he worked on prosthetics, railway signalling and traction; he was active in the British Association for the Advancement of Science, campaigned for education, helped to found the Yorkshire Philosophical Society (1821), and was Member of Parliament for Scarborough and a political reformer.

Another friend and parliamentary colleague of Cayley was his younger neighbour, Sir John Johnstone of Hackness Hall. We have met Johnstone before: he was the patron and employer of William Smith, first delineator of the geological strata of England and Wales. Thus we find two of the world's most influential scientific minds in adjoining Yorkshire townships at the same time.

Herr Magersuppe landed the innovative Tern glider on Scarborough's North Beach, not far from William Smith's Rotunda museum, 130 years after Cayley demonstrated that such flight was possible. Tiltman, Johnson, Boultbee and the others are like threads, each an individual self yet spinning out of Cayley's original insights to make a weave criss-crossing Yorkshire from Saltburn to the Humber. Among them, like a streak of colour, is Sarah Burton, her 'astonishing' red hair in the slipstream over Bridlington on a clear May morning.

9

A WET DAY IN BAWTRY

━━━━━━━━━━

The parish church of All Saints, North Street, stands by the River Ouse in downtown York. It is set a few yards back from the river's west bank, and contains one of the most extraordinary medieval stained-glass windows in Europe. The window was made in about 1410–20 at the expense of members of two of York's leading families, some of whom are depicted kneeling in the lowest row of panels.[1] And kneel they well might, for what the window shows are portents of a burned-out world.

The omens had been described in a poem called 'The Prick of Conscience', which was written in Yorkshire around 1340 and had since circulated widely.[2] The signs were expected on fifteen successive days, each of which is depicted in its own panel. The scenes are annotated by captions that paraphrase verses in the poem. On day one the sea rises to the height of mountains; on day two it recedes, and on day three it returns to its usual level. Fishes in the sea make roaring noises on the fourth day; on the fifth, the sea burns. Over the next four days, trees and grass bleed, buildings fall, rocks and stones bang together, and there are earthquakes. Rough places are made plain on the tenth day, and on the eleventh people who have been in hiding emerge speechless because they are in shock. The twelfth sign sees bones of the dead coming back together; duly reassembled, the dead rise and stand on their graves. On the thirteenth day, stars fall from heaven (Plate 19).[3] All living things die on day fourteen, preparatory to judgement. On the fifteenth day 'the world shall burn'.[4]

I first heard about this countdown to Armageddon in 1971. The church was ten minutes' walk from our office. One autumn afternoon I walked into the north aisle and came face to face with the donors. They looked richly clad, well fed and worried. I was struck by their faces: rounded, wide-eyed, lips slightly parted. Their opened mouths suggest gasps of alarm, but surely connote praying aloud.[5] Above them, anxious bystanders peered out from panels coloured by ruby reds, amber and gold, different blues – azure, cobalt, lapis. Higher still, lit by the October sun, falling stars trailed streaks of flaxen fire, until they hit the ground and shattered in radiate flashes.

In 1971 these scenes had an oddly contemporary feel. Nine years before, the world had been on the brink of nuclear war over the introduction of nuclear missiles into Cuba. For a few days in October 1962 there was a real sense that the earth might imminently flare and life be gone. Since then the superpowers had probed, parried and challenged each other through proxy wars around the world in which millions had died. If any one spot epitomised the remark of Sir Michael Palliser, a former permanent Under-Secretary of State at the Foreign Office, that the Cold War was 'a thread that has run through everything',[6] it was a place not far from All Saints, North Street, about which everyone knew yet was not on the map.

As already noted, Yorkshire people love Whitby and because of that they love the ride to get there. At first there are familiar waypoints that tell you the coast is coming closer – a roadside café, that turning, a knot of trees on the skyline. Expectancy builds as you leave Pickering and the road climbs through the Tabular Hills onto the North York Moors. After seven miles, a surprise: to your left the earth vanishes into a hollow 400 feet deep and three-quarters of a mile across. This is the Hole of Horcum. If you stop here to take things in, the surrounding moors resolve as a succession of differently shaded distances in which far-off nabs and plateaus peep over the shoulders of those nearer.

The Hole of Horcum is marinated in fable: the Hole was dug by a giant; the hairpin bend down which you go when you carry on towards Whitby is the Devil's Elbow; in the Saltgergate Inn below was once a peat fire that burned for 200 years.[7] But none of that prepares you for what follows: a truncated polyhedron standing alongside some low buildings about a thousand yards to the east. The structure looks a bit like a vast loudspeaker, and in a world rearranged by giants it could doubtless have been plonked down by some passing audio engineer. In fact, it is a radar station. Its name is RAF Fylingdales, and it has siblings in Greenland and Alaska. Their joint purpose is to warn the United States and UK of missile attack, to follow orbiting satellites and keep track of man-made objects in space.

Fylingdales came into use in 1963. There were three tracker radars, each weighing about 112 tons and steered by machinery. Three scanners were used to provide complementary data. Two of them searched from side to side at slightly different elevations just above the horizon. A target appearing in both registers would potentially be a missile accelerating under its own power. Such a target would be tracked by the third radar to ascertain its trajectory and point of impact. Today all that sounds very clunky, but they worked almost without pause for nearly three decades. The scanners were enclosed by spheres and so invisible to the public. The spheres, in contrast, were very obvious indeed (Plate 21). They were taller than York Minster's nave and regardless of their omission from Ordnance Survey maps on security grounds they could be seen for miles. People rather took to them, coming to see 'the golf balls' more as an extension of Yorkshire's landscape than as intrusions into a national park. For a time they even became a tourist attraction, and a bridge between fiction and fact. The launch of *Dr Who* coincided with the station's opening, while Cold War anxieties, Yuri Gagarin's orbit of the earth in 1961 and rising TV ownership encouraged a mass audience for the BBC's seven-part science fiction drama *A for Andromeda* (1961). The series was set ten years in the future, in 1971, and co-written by the Yorkshire-born cosmologist Fred Hoyle. One of its purported locations was a new radio telescope in Yorkshire at a place called 'Bouldershaw Fell'. Supercomputing, genetic experiment and orbital missiles were entwined in the story. Nearly 13 million people watched its last episode.

Fylingdales in its turn looked east. We are about to see that there is a long tradition of doing so, and that Whitby is a good place to introduce it. But first let us savour the journey. The moors are clad in bracken, cotton grass, bilberry and heather, and your experience is coloured according to the season in which you make it. Mocha, russet and dark greys surround you in winter. In spring the road is wreathed by greens and flecks of yellow. Late summer is purple, when tiny heather flowers emerge in their billions, sweetening the air until they gradually fade to a leached pink, then darken as the year turns.

Your first sight of Whitby is from Blue Bank, in the euphoniously named parish of Eskdaleside cum Ugglebarnby. This is where the A169 topples into a one-in-four descent, and there ahead, silhouetted against the sea, are the outlines of the parish church and abbey, poised on East Cliff above the red-roofed town around the mouth of the River Esk. We have seen that the abbey had its beginnings in the seventh century under the leadership of Hild, the daughter of a Deiran prince.[8] The building you see now is more recent. So is the place name. Whitby is first recorded late in the eleventh century, and since the name is of Scandinavian origin it was then unlikely to have been more than one or two hundred years old. The eighth-century historian Bede knew the monastery by its previous Old English name, *Streanæshalch*.[9] The meaning of that name is a puzzle, as is Bede's explanation of it in Latin as *sinus fari*.[10] *Sinus fari* is usually translated as 'bay of the lighthouse', but whatever Streanæshalch meant it cannot have been this. Was Bede writing figuratively? Peter Hunter Blair (1912–82) pointed out that *sinus* typically denotes a crease or fold of a garment and that *farus (pharus)* is often used in the sense of a lamp or candelabra, and by extension a source of brilliance. Hunter Blair further noted that Bede's wording shows that he was 'not here translating an English name into a Latin equivalent' but rather interpreting it. The idea of a 'bay of light' points us towards a later passage in the *Ecclesiastical History* where Hild's mother dreams of the abduction of her husband. Searching for him, she finds a precious necklace hidden under a garment. The necklace – like the jewellery worn by the princess at Loftus – seems to spread 'a blaze of light'. The dream, said Bede, allegorises the birth and example of her daughter Hild.[11]

Part of the case for this alluring explanation was the lack of a lighthouse. If there had been one, reasoned Hunter Blair, it would have been Roman, and so sizeable, and we would know about it. But did he reckon with the greed of the sea? Cliff recession since Hild's day has been of the order of a third of a mile.[12] A Roman lighthouse could well have been and gone.

Another building that could have stood on the vanished headland belonged to a family of Roman military structures known as *burgi*. *Burgi* were observation towers, sometimes enclosed by defensive outworks, with upper galleries or belfry-like top storeys to provide good all-round outlook. They were built in various forms and sizes and are normally thought to have made up networks whereby news of raids could be collected and rapidly shared over long distances through signalling by fire, smoke or flags. Second-century inscriptions show that their purpose on the Rhine-Danube frontier was to counter infiltration by groups of raiders.[13]

A coastal chain of observation posts was put in place by the Romans between Huntcliff and Filey in the later fourth century. Five sites survive in whole or part (at Huntcliff, Goldsborough, Ravenscar, Scarborough and Filey) and two more are suspected to have existed at Hartlepool and Flamborough. The series is thus likely to have stretched for sixty miles between the Tees estuary and Bridlington Bay. Contemporary terminology for them is shown by a dedication slab from Ravenscar:

'Justinianus, commander; Vindicianus, magister, built this tower and fort from ground-level'.[14]

Excavation shows that the towers were around 45 feet square at the base. This scale, and the geometry of visibility between stations, suggests that they rose at least about 100 feet and possibly higher.[15] However, there are no lines of sight between Huntcliff and Goldsborough or Goldsborough and Ravenscar. Hence, if the Yorkshire stations functioned as a system, two of its members must be missing. Coastal topography indicates where they would have stood: at Boulby (close to the later resting place of the Loftus princess) and Whitby.

The towers have a likely historical context. Sources speak of crisis in Britannia in 367-8, involving internal breakdown and incursions by Scotti, Picts and Saxons. These reports imply some degree of coordination between Saxons and Picts along the length of the east coast. One account of the emergency records the dispatch to Britannia in 368 of a senior officer, Flavius Theodosius, who restored stability, reorganised units, and established new frontier outposts.[16] Such events offer a ready context for the towers, but maritime marauding continued and the observation posts could just as well have been put in place to counter later raids.[17, 18]

Independent evidence for the stations' use is limited. Three sites (Huntcliff, Scarborough and Filey) have been cut by the sea (Fig. 42), and where excavation has taken place it was usually in the nineteenth and earlier twentieth centuries when enthusiasm outstripped technical skill.[19] Only Filey has benefited from modern excavation.[20] Here, analysis of food remains, pottery and coins points to supply and remuneration 'controlled from some regional centre, which continued to function into the last decade of the fourth century, if not into the first decade of the fifth'.[21]

It is a question how the system was meant to work. If its primary purpose was to pass intelligence to the Roman high command one would expect relay stations between the coast and bases at Malton and York, yet no sites of this kind have yet been found. It is possible that the towers were intended to work in conjunction with vessels of the Roman fleet, for example by alerting crews to the approach of potentially hostile craft that would not be visible at sea level. A further possibility is that the towers were positioned to give warning of imminent landings in specific places. If the aggressors' boats were in the structural tradition of vessels like the early fourth-century 75-foot Nydam boat – clinker-built, of shallow draught with high prow and stern – then flat beaches

42. Site of Roman signal station at Scarborough

were the places to put raiding parties ashore. All but one of the known stations overlooked such beaches.[22]

Whatever the system was, the Filey excavation gives a haunting glimpse of how it ended. Upper deposits of the courtyard contained concentrations of bones of small mammals like field and harvest mice, voles and shrews. The remains derived from owl pellets, most likely from barn owls or short-eared owls. Barn owls are mistrustful of human presence, preferring abandoned buildings in which to roost, while short-eared owls are even more wary of human intrusion.[23] The owls point to a last phase early in the fifth century when the signal station was vacant or only intermittently used.[24]

The owls were hunting around the same time as the first of several thousands of burials were being made on Sancton Wold, about 30 miles to the south-west. The dead were cremated, and their crumbs of calcined bone – sometimes comingled with those of animals, and accompanied by objects – were interred in hand-made pots in a way that suggests a close relationship with the region we now call Schleswig Holstein and around the mouth of the River Weser.[25] Whether these people were members of an entire community that translocated across the North Sea, or whether they were there by invitation – for instance to counter coastal raiders under the auspices of whatever was left of the military command in the Malton region after the Roman regular army withdrew – they invite us to look east, and north.[26]

So do the owls. Some short-eared owls are winter visitors from Scandinavia, while recoveries of Scandinavian owls that have been rung show that they may travel as far as Russia and the Mediterranean. The range of this long-winged, charismatic bird is more or less coincident with the arc of Yorkshire's trade, which in the Viking age extended through the Gulf of Finland along Russian rivers to the Black and Caspian Seas, and so into Asia. Commodities reaching the west included sable furs, amber, honey, wax, pitch, and silver. A lot of silver: by the early eighth century more silver coin was circulating in eastern England than at any time in the next five hundred years. Much of the new silver came from the east. It is not clear what was going the other way. Specialised textiles (like heavy cloaks) and grain are mentioned in earlier sources, and it may be that there were certain key commodities like slaves that do not show in the archaeological record. By the tenth century there was an expanding traffic in wool with Frisia and Germany that anticipated the wool trade of the later Middle Ages and drew in silver from new-found sources in the Harz Mountains.[27] Silver was also available in smaller quantities in parts of Yorkshire and nearby areas. Silver occurs as a constituent of lead in the veins that were formed by injection of hot mineralising solutions into fissures and cracks in Carboniferous limestone of the Peak and upper Dales sixteen million years ago.[28] Some have supposed that these deposits were too meagre to have been worth refining, but it is clear from late medieval sources that they were so worked, while in the eighth century Bede described Britain as a place with 'rich veins of metal, copper, iron, lead, and silver'.[29]

Yorkshire's eastern horizons are typified by a hoard buried in the Vale of York in the late 920s which included coins minted as far away as Baghdad, Uzbekistan and Afghanistan.[30] Four hundred years later grain and wool were shipped to the continent and Hanseatic warehouses operated in York and Hull. Here is the context for elite York families like the Henrysons and Hessles who gave generously for new windows to adorn their parish churches.

Here, too, is the context for the strategic and economic importance of later medieval and early modern Hull, and the recurrent efforts made to protect it. Until the later eighteenth century the River Hull itself was the anchorage, and the town stood in the western angle between the Hull and Humber. During the fourteenth century, the western limits of

the town were enclosed by a wall, with the River Hull acting as a moat
on the east. John Speed's map of 1611 shows the town bordered by a high
wall punctuated by large towered gates, posterns and interval towers.
To the east of the River Hull a castle, ditch, rampart and wall running
between blockhouses were added in the sixteenth century. More works
ensued during the Civil War, and again during the Anglo-Dutch wars
that followed. A visitor in the 1670s described the landward defences:

> Being now got to it and ready to enter, we saw a draw-bridge and a
> broad and deep moat full of water surrounding this part of the town;
> leaving this behind us we came to another deep moat of water with
> a drawbridge over it where there is a strong gate-house, gates, and
> portcullis, and a strong wall on the inner bank surrounding this moat.
> Then allowing room for defence where men may stand and use their
> arms we came at length to another strong gate which let us into the
> town, with a wall surrounding their houses, both walls and gate-
> houses being well stored with guns to annoy the enemy whenever he
> shall come . . .[31]

And to seaward:

> Upon the northern side of the town lies the haven in which a great
> many ships may ride; at the entrance is a great chain to keep out
> intruders when they please, and on the other side of this haven here
> is a strong wall, and at the end to seaward a good castle well planted
> with guns, and another fort some half a mile off . . .[32]

Defoe noted airily that since Hull lay open to the sea it was vulnerable
to bombardment, but that could be prevented 'by being masters at sea,
and while we are so, there's no need of fortifications at all; and so there's
an end of argument upon that subject'.[33]

Defoe did not bargain with aerial bombardment, and when the Great
War began no one could describe Britain as masters of the air. In 1909
the Committee of Imperial Defence had concluded that aerial attack on
any scale was unlikely in the foreseeable future and that public spending
in this area should be limited to a single experimental rigid airship. This
was commissioned by the Admiralty, whose interest faltered after the
airship broke in two before it flew. In 1913 Winston Churchill (First
Lord of the Admiralty) voiced growing unease when he said that it
would be advisable for Britain to make a start in building two rigid
airships 'so that the art of making them is not wholly unknown to us'.[34]
In Germany, by contrast, enthusiasm for flight had been growing since

the later nineteenth century, witnessed in the growth of balloon clubs, military ballooning and experiments with the rigid airship. Trials of airships in the early 1900s by the Zeppelin, Parseval and Schütte-Lanz companies attracted wide public interest. By 1914 commercial Zeppelins had carried over 10,000 passengers on pleasure flights and special voyages. They now turned to carrying bombs. On 19 January 1915 three Zeppelins set forth from a base near Hamburg to bomb targets around the Humber. One turned back and the others experienced navigational difficulties that led them to attack places in East Anglia. But the writing was on the wall: this was the first of fifty-one Zeppelin raids against Britain, eight of them on Hull.

Defences against the new threat were at first slight and ineffectual, but by 1916 there was a national organisation that comprised gun sites, searchlights, fighter defence, and a system for observation and tracking. One of the anti-aircraft commands centred on the Tees (which had a dependency at Skinningrove) and another on the Humber. The nerve ends of the system were listening devices to give advance warning of inbound Zeppelins. These included large parabaloid concrete mirrors that were designed to give a few minutes' early warning by concentrating the sound of approaching engines at a focal point, where it was collected by a funnel-shaped instrument akin to an ear trumpet (later a microphone) and then assessed by a duty operator using a stethoscope.[35]

Two of the Yorkshire sound mirrors still exist.[36] One is at Boulby, close to the burial place of the Loftus princess and probably the site of a Roman observation tower. The structure is an upright shallow dish of cast concrete. It stands about sixteen feet high, has a slight backward lean and is flanked by blinker-like walls that were intended to screen out ambient sound. The mirror was oriented to cover approaches to the Tees, Redcar and Skinningrove. In 1916 it was isolated; today it is in the middle of a housing estate and looks a bit like a ruined bus shelter.

The other mirror stands alone, like a prehistoric monument, in a large arable field on the Holderness coast near Kilnsey (Fig. 43). In heft and stance it reminds you of an Easter Island statue with no face. Yellow-orange lichen marks the passing of a century. The concrete ear hearkens to lost sounds yet is deaf to everyday winds. It is one of Yorkshire's strangest and most affecting structures, and radar made it obsolete over-night. By 1939 a chain of south- and eastward-looking radar stations ran from Southampton to the Orkneys. By 1943 the system had been extended and upgraded with radars of greater long-sightedness and acuity. Like some of the sound mirrors, a few of them were put in places

43. Acoustic mirror, Kilnsea

that coincided with the positions of east-watching Roman precursors.[37]

Here is an unlooked-for link with another stained-glass window. One reason why radar existed in 1962, or in 1950, when Ted Hughes whiled away his hours staring at one during National Service at Patrington, or in 1940, when more than any other invention it helped stave off defeat, was the foresight and drive of a Yorkshireman called Philip Cunliffe-Lister. In 1935 Cunliffe-Lister was appointed Secretary of State for Air. The concept of radar was demonstrated by Robert Watson-Watt, but it was Cunliffe-Lister who, in tandem with Hugh Dowding, insisted both that the RAF should acquire the newly designed Spitfire, and that its use should be controlled by an air defence system based on radar (Fig. 44). Both policies offended then-prevailing airpower doctrine but Cunliffe-Lister drove them through. In 1938, Cunliffe-Lister, now Lord Swinton, resigned following the failure of Supermarine and their parent company to meet Spitfire production deadlines.

Swinton ... Swinton Park, just north of Ripon, now a hotel, was his seat, inherited through his marriage to the granddaughter of a Bradford industrialist whom we shall meet later. They worshipped in the parish church at Masham, and if you go there you will see a memorial to their sons.[38] It is a window. Its designer was Harry Stammers, a London-born artist who set up at glass-painting studio in York in 1947 and over the next twenty years produced some 180 stained-glass windows. In

*44. Looking east in 1940: Chain Home radar station on Danby Beacon.
Danby was one of the parishes which saw substantial nineteenth-century
emigration to Canada (p. 102).*

Stammers's windows you see past and present side by side: medieval
saints, bishops and angels; 1950s families, workmen, typists and miners;
and a love of stars and nature. Stammers loved birds. You will see his
work in a number of cathedrals,[39] and it is well represented in church
windows down Yorkshire's east coast, at Eastrington, Withernsea, Sunk
Island, Hedon and Hull.[40]

His window at Masham depicted episodes in the lives of two saints
(Plate 20). One is a St George, who wears white armour and stares with
concentrated intent along the shaft of his spear, which is aimed at the
breast of the dragon. The dragon is realised in rich reds and ruby. It
confronts George in a semi-heraldic stance, advancing, right front leg
lifted, head thrown back in a Guernican pose to snarl up at the sky, with
wings that might be mechanical. To the rear in blue shadow is a terrified
princess, peering past lifted arms to know the outcome.

The other is St Francis, of whom a story is told of his preaching to
bystanders. He spread his arms to attract birds. Though he preached
in Italian, hearers of other nationalities heard his words in their own
tongues. Some attributed this to the birds, each bird relaying the words
in a different language. The birds here are from Yorkshire (Plate 22).
Among them are grouse, pigeon, lapwing, a heron from the Ure, the
curlew that calls in spring, a raven, a peacock no doubt from the grounds

of Swinton Hall, its tail modestly folded. Some of the birds wear hints of expression – a crow's beak is parted in a call back to the speaker. Between them are splashes of colour, and stylised poppies.[41] At Masham, is a moorland Pentecost.

Only nineteen years separated Ted Hughes's sojourn at Patrington from the Cuban Missile Crisis of October 1963. On 27 October a US reconnaissance aircraft was shot down over Cuba, another US aircraft inadvertently overflew the Soviet Union, and the US Navy came close to provoking a Soviet submarine into launching a nuclear-armed torpedo.[42] The Pentagon remembers that day as 'Black Saturday'; Yorkshire remembers it as grey and wet. At 1 p.m. Thor missiles stationed near Howden, in Holderness and on the Wolds were put on fifteen minutes' notice of launch. Their targets included places with which York and Hull had traded for hundreds of years, and to which budget airlines now fly from Doncaster Robin Hood. Two days later fifty-nine of the UK's sixty Thor missiles were poised for release 'simply by use of the telephone'.[43]

The man in immediate control of these weapons was Sir Kenneth Cross, Commander-in-Chief of Bomber Command, who in certain circumstances was able to authorise their use. In the event of war and breakdown of communication his authority would pass to regional deputies. In the case of Lincolnshire's bombers this meant Air Vice-Marshal Sir Patrick Dunn, whose headquarters in Bawtry Hall, a redbrick eighteenth-century family house, was just across the county boundary in Yorkshire.

So it was that after 2,000 years of looking east and north from Yorkshire, the world on that rainy Saturday was one phone call away from a time when rocks and stones shall strike together, and short-eared owls will come to roost in the ruins of Bawtry Hall. But the telephone did not ring and the Pricke of Conscience window in All Saints, North Street, is still there, performing its original duty of alerting visitors to what will happen at the end of time. Go and look. Go while you can. Better, go when the glass is lit by October sun, and stones melt as falling stars hit the ground.

WEST

The West Riding was the largest and most populous of Yorkshire's three historical divisions. It spanned the Pennines from the Humberhead plain in the south-east to the threshold of the Lake District in the north-west, and in 1901 over three quarters of Yorkshire's 3.59 million people lived there, most of them in an agglomeration of industrial cities and towns east of a line between Skipton and Sheffield.

Many of these places were only a tram-ride apart. They thus liked to celebrate their separateness, and found many ways to do so. Sport was one: the West Riding was the birthplace of club football and rugby league, and communities competed on the cricket pitch at least from the eighteenth century.[1] Architecture was another, through the individuality of chapels, railway stations, arcades, mechanics' institutes, wool and corn exchanges, or social modernism of the kind seen in Leeds' Quarry Hill Flats (1934–41) or Park Hill (1957–61) in Sheffield. Music provided a further medium for expression of local selfhood. Collieries, mills and townships fostered brass bands. Many towns had their own choirs and municipal orchestras, and spacious, colonnaded, domed town halls in which they performed. Choral classics like Vaughan Williams's *A Sea Symphony* (1910) and Walton's *Belshazzar's Feast* (1931) were first heard in Leeds town hall; so also Elgar's symphonic poem *Falstaff* (1913), while since 1951 Arthur Wood's 'Barwick Green' has been heard six days a week as the theme tune for *The Archers*. Wood (1875–1953) was born in the Calderdale blanket-making town of Heckmondwyke; the Barwick he had in mind was Barwick-in-Elmet, a village east of Leeds known locally for its maypole, and nationally for its evocation of the lost post-Roman kingdom to which the West Riding may be heir.

Many of Yorkshire's best-known natural features and cultural monuments lay in the old West Riding. Examples are Malham's limestone cliffs and pavements, the Three Peaks, the Ribblehead Viaduct, the Brontë parsonage at Haworth, Halifax's Piece Hall, the stupendous mansion at Wentworth Woodhouse built by the Marquesses of Rockingham, and two World Heritage Sites: the Cistercian abbey of Fountains, and Saltaire's industrial village.

The West Riding's breadth and self-sufficiency made a nursery for creators like Alan Bennett, Barbara Hepworth, David Hockney, Ted Hughes and Henry Moore. It also had something to do with a liking for proverbial sayings that tease outsiders by presenting opinions that might be real, self-mocking, or both. For instance:

> 'See all, hear all, say nowt;
> Eat all, sup all, pay nowt.
> An' if ivver tha does owt fer nowt –
> Allus do it fer thissen.'

Or a Leeds Loiner's advice to his son:[2]

> Let thi eye be thi judge,
> Thi pocket be thi guide, an'
> Brass t'last thing tha parts wi'.

Many of those who live within the compass of the old West Riding still think of it as their home. However, the administrative county that it became in 1889 was shattered by the reorganisation of 1974, when the shards were distributed between seven local government areas,[3] several of which have themselves since been destroyed.

10

IN BARNSDALE FOREST

B arnsdale Bar is on the A1 between Doncaster and Pontefract. The 'bar' recalls a tollgate a short distance to the south where the Great North Road and Pontefract Road once met. Today this is an elaborate interchange, whence roads sprout off to Pontefract, nearby quarries, business parks, and a place called Campsall to which we'll come. For passing A1 drivers Barnsdale Bar is marked by a service area on the southbound side, and a Pulse and Cocktails Adult Superstore if you're headed north. The superstore is on the site of a Little Chef that many will remember.[1] But aside from that place mat, the shiny menu offering the Olympic Breakfast and the lollipop on departure, how many customers realised that they were sitting within a couple of longbow shots of the Stone of Robin Hood?

Robin Hood's Well, Robin Hood's Cave, Doncaster Robin Hood – there are Robin Hood names all over the country, and if you study them you will find that most were conferred in the nineteenth century and that about a third of them are pubs. However, a few are older, and a handful, of which this is one, are very old indeed. Barnsdale's Stone of Robin Hood is mentioned as a boundary marker in a land grant of 1422.[2] That might be no more or less suggestive than the track or road called 'Robhodway' recorded in Suffolk in 1344,[3] or John Leland's account of the 'fisher townlet of 20 boats called Robyn Hudde's Bay' in about 1535,[4] were it not for one thing: many of the earliest surviving references to and stories about Robin Hood put him in Barnsdale.

'Robehod' was well known as an outlaw personality by the middle of the thirteenth century.[5] No one knows whether anyone real stood behind that reputation, or when and how stories began to circulate, but over the next two centuries 'Robehod' evolved into 'Robin Hood', who became a compound of different people and incidents. Thus emerged the legend of a 'forest outlaw who defied the law and still managed to remain free for many years'.[6] The legend ballooned, hybridised with other tales, and has continued to soak up new material ever since.[7]

The first stories to survive in manuscript date from around 1450.[8] By then, the public knew Robin as 'a stock character on which different adventures were hung'.[9] The stories have been likened to episodes in long-running adventure series, like Superman or Batman, wherein a hero is involved in a succession of exploits 'against the same set of villains in an infinitely changing set of circumstances'.[10] They were known as 'rymes' and there are signs that more of them circulated than have come down to us. The rymes that do survive contain a core of features that in some respects differ from those known to today's cinema-goers. Robin Hood is a yeoman, for instance, not a cast-out Anglo-Saxon earl. The period is later – no Norman yoke – and Robin does not rob the rich to give to the poor. Maid Marian and Friar Tuck, prominent in later iterations, are not there. Neither is Prince John.

Other elements are common both to the early stories and to later variations. Robin's habitat is the greenwood. There are merry men, and among them are Little John, Will Scarlok and Much (whose introduction as 'Much the miller's son' became misremembered as 'Much the Miller' in later retellings). They do hold up travellers and sometimes offer hospitality to their victims. Paradoxically they are honest robbers: they are for the common good and against self-interest in high authority. Their anti-clericalism is selective, targeting venal abbots and bishops but not sincere parish clergy. Robin himself is devout; he has his own chapel and would have stood up for Chaucer's Poor Parson of a Town. Their persistent adversary is the Sheriff of Nottingham, in and near whose castle several exploits occur. They hunt the king's deer but profess loyalty to the king himself, who in these stories is King Edward. But which Edward is a question.

Sometime in the later fifteenth century five of the Robin Hood tales were brought together in one gathering called *A Gest of Robyn Hode*.[11] The *Gest* was organised in sections called *fyttes*,[12] and furnished with a storyline that connected the different rymes in a more-or-less logical sequence. The collection caught the eye of different printers, who between them produced at least five editions between the 1490s and 1550.

It is in effect a minstrel's serial, designed to be recited at intervals. It includes what is perhaps the earliest story of all, the tale of the impoverished knight. In this story, Robin assists a knight who has mortgaged his lands to the abbot of St Mary's, York, by robbing the monks themselves to repay the loan. The knight later becomes Sir

Richard of the Lee who fortifies his castle to protect Robin and his
men from the vengeful sheriff.[13]

Robin's heartland in the *Gest* is Barnsdale. At the start of the first section:

> Robyn stode in Bernesdale,
> And lenyd him to a tre.[14]

Towards the end of the eighth section, now in royal service after a
pardon, Robin 'longeth sore for Bernysdale':[15]

> Forth than went Robyn Hode
> Tyll he came to our kynge:
> 'My lorde the kynge of Englonde,
> Graunte me myn askynge.
> 'I made a chapell in Bernysdale,
> That semely is to se,
> It is of Mary Magdaleyne,
> And thereto wolde I be.'

Robin is granted leave to go back for a limited term. He reaches the
'grene wode' in a 'mery mornynge' and listens to birdsong.

> 'It is ferre gone,' sayd Robyn,
> 'That I was last here . . .'

Then he gives a joyous blast on his horn so that all the outlaws of the
forest should know that he is back. Robin does not rejoin the king;
instead, he returns to his old ways and lives on in Barnsdale for another
twenty-two years. Having flouted the king 'he knows there is no going
back, no second pardon'.[16]

Between the *Gest*'s opening and the homecoming, Barnsdale is the
repeated place of reference. When Robin and Little John resolve to
waylay a traveller, 'they looked in to Barnsdale'.[17] This is where they meet
the downcast knight and embark on the project to restore his fortunes.
In the second section the knight comes back and pauses to watch wrest-
ling at Wentbridge (Fig. 45).[18] Later, having 'looked in Barnsdale by the
highway', Robin and Little John intercept two Benedictine monks.[19] We
have been here before, of course. The highway is the Roman road that
runs along the limestone ridge north of Doncaster, and Wentbridge is
at or close to the place where the Northumbrian king Oswiu destroyed
the forces of Penda and his allies at Winwæd in November 655, and very
likely in the vicinity of Brunanburh in 937.[20]

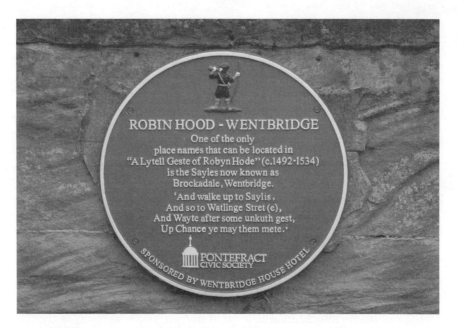

45. Plaque at Wentbridge

Historians warn against taking a literal view of Barnsdale or Sherwood.[21] Robin Hood's greenwood, like 'the north', was imagined and
idealised. Barnsdale in the *Gest* is a literary stereotype where it is always
early summer and whence Robin and his always-merry companions can
be teleported in and out of Nottingham without slogging through the
forty miles between.[22] Here and in other early tales the references to
place take it for granted that listeners or readers will lack both local
knowledge and the need for it.[23] Like arcadia, Cockaigne, or the big
rock-candy mountain, Robin's greenwood is a type of ideal society that
is at once anywhere and nowhere.[24]

If this is right, and surely it is, some puzzles remain. If the greenwood
in the *Gest* is 'a strictly literary locale', how comes it that so much of its
subsidiary detail maps directly onto Barnsdale and its surroundings? To
give examples, when Robin and Little John go in search of their first
quarry, they 'walk up to the Saylis, and so to Watling Street'.[25] For some
years Robin Hood scholars have noted that 'the Saylis' is an area close to
Wentbridge in the adjoining township of Kirk Smeaton.[26] In the ballad
'Robin Hood and the Potter' Little John waylays the potter and his cart
at *Went breg*.[27] This length of what became the Great North Road is the
highway providing a north–south passage between marshland and ever-
rising hills north of Doncaster: just the place for an interception (Fig.

46).[28] Such particulars seem needless in a genre of literary convention.

More examples reinforce an impression that the imagined green-wood has been draped over a real district. The dedication of the local parish church at Campsall – Mary Magdalene – coincides with that of Robin's personal chapel.[29] When Robin and Little John (whom the *Gest* says comes from Holderness in the East Riding) meet the downhearted knight, the knight is aiming to reach Doncaster or Blyth within the day.[30] That would be a realistic journey, and there are places in Barnsdale from where you can actually see Blyth on the southern horizon. Other journeys are made to nearby places, like York. In the seventh section the king searches for Robin:

> All the compass of Lancashire
> He went both ferre and nere,
> Tyll he came to Plomton Parke
> He faylyd many of his dere.[31]

The excursion into Lancashire could illustrate the contention that nei-ther compiler nor reader of the *Gest* was concerned with geographical accuracy, or a possibility that the *Gest* is a composite that drew from different cycles of rymes. In the first case it would not be necessary to look, as some have, for a Plompton Park in Lancashire or Cumberland when there is one close to hand. Just off the highway going north, two

46. Milepost on Great North Road, Wentbridge

miles short of Knaresborough, was the manor of Plompton, in the midst of Plompton Park.[32]

At the end of the *Gest* the 'wycked' prioress of 'Kyrkely' conspires with her lover 'Syr Roger of Donkestere' to slay Robin. Sir Roger is clearly local. 'Kyrkely' is generally taken as Kirklees, near Brighouse, where there was a Cistercian nunnery. That was the understanding in the sixteenth century when John Leland recorded that Kirklees was the place where the 'noble outlaw' was buried (*ubi Ro: Hood nobilis ille exlex sepultus*).[33, 34]

What was Barnsdale? These days it is often talked of as 'Barnsdale Forest', and there has been speculation that the forest originally covered a large area. Some have gone as far as to suggest that it once stretched south beyond Doncaster to merge with Sherwood.[35] Today's Ordnance Survey simply prints 'Barnsdale' across a district of smooth hills and dry valleys either side of the A1, just south of the Barnsdale Bar service area. Reasonably enough, given all the greenwood references, many commentators have supposed it originally to have been covered by woodland. If so, however, there was a change of land use earlier in the Middle Ages, for archaeology finds that in later prehistory and through the Roman period this was a predominantly agricultural landscape of enclosures, trackways and fields.[36]

Such an area would not necessarily by inconsistent with a forest, which in the Middle Ages was a legal idea rather than a particular kind of scenery. Forest could be any kind of landscape – it did not equate with trees. The word derives from Latin *foris*, 'outside', in the sense of land outside normal jurisdiction. A forest was land set apart where deer and other game could be conserved for hunting by kings and magnates. This legal exceptionalism meant that people needed to know where forest borders ran. Real Yorkshire forests like Galtres and Knaresborough were accordingly described in detail and later mapped, while their potential for franchises, payments of tenancy and leasing, together with their workings as administrative units, generated a network of courts and tiers of specialised officials. It is odd, then, that no district called Barnsdale Forest, nor forest eyre, warden, verderers, regarders or any of the other administrators involved in forest management, show up hereabouts in medieval or early modern sources.

It is odd, too, that visitors in the sixteenth and seventeenth centuries found Barnsdale in different places. The map in Drayton's *Poly-Olbion* (1621) shows Barnsdale out on Yorkshire's edge, about twelve miles away to the south of Hatfield Chase and next to Rotherham.[37] William

Hole, Drayton's engraver, used a bushy tree convention for forests and chases like Galtres and Hatfield. He did not use it for 'Barnesdale'. John Leland described a low-lying area framed by Cawood, Selby and the River Aire ten miles to the north, where he 'saw the wooddi and famose forest of Barnesdale, wher they say Robyn Hudde lyvid like an owtlaw'.[38] Today the area is open, drained, wide-skied and ploughed; in 1841 the Old Series of the Ordnance Survey recorded local names like Ling Wood, Spark Hagg, Boggart Bridge, Wood Ends Farm and Melton Leys, which recall a time when the district consisted of woodland and copses interspersed with clearings, heather heathland and peat marsh. The largest of the woods appears on Christopher Saxton's map of 1577, where it was named as 'The Owt Wood' – not Barnsdale and not a forest.[39] What this means is that by the mid-sixteenth century some local people had shifted the site to the nearest sizeable 'wooddi' area to correspond with the descriptions in the poem. Another puzzle is the 'dale' in Barnsdale. If 'dale' means 'valley', where is it? The nearby valley of the River Went is already spoken for, and the Skell brook that trickles between Skelbrooke and Skellow barely qualifies.

An area that is elusive and topographically perplexing would go well with a greenwood of imaginative convention. It would also fit with the medieval perception of forest as a proxy for wilderness, somewhere on the edge of things that symbolised both renewal and refuge for those who lived or worked on the fringes of society, mediating between desert and culture, like hermits and huntsmen. Or outlaws.[40] And there we could leave it, but for some further facts.

A building called Barnsdale House is shown beside the road from Barnsdale Bar to Campsall on the Old Series of the Ordnance Survey.[41] Three fields to the south is a hilltop folly called Barnsdale Summer House, designed in 1784. From here there are long views: you can see far to the east, and south into Nottinghamshire. Much of the land in this locality was marginal. A recently wooded area behind Barnsdale House, for instance, was known as New Whin Covert. 'Whin' is a word for gorse, a prickly pioneer shrub that colonises unpromising ground. Other whin names are nearby – Whinny Hill, Barnsdale Whin, Whin Bed – and the sense of marginality is confirmed by a record of 1738 providing for a warren upon 'Barnsdale Waste' in the manor of Campsall. The building on the site of Barnsdale House today is called Warren House Farm, while another eighteenth-century land transaction mentions Barnsdale as a piece of land adjoining the parcel concerned.[42] What these sources show is that Barnsdale was neither large nor vaguely defined: rather, it was a distinct hilltop and hillside neighbourhood on the borders of the townships of Campsall, Skelbrooke, Kirk Smeaton, and North Elmsall.[43]

Could the hill explain the dale? Analysts of place names work forwards from their earliest written forms. The first local spelling we have for Barnsdale concerns a strip of land called *Barnysdale ryg* in a land transaction of Monk Bretton Priory in 1462.[44] This has been ascertained as 'Beorn's valley', which is entirely reasonable. However, it is often the case that early forms of later identical names lead to different original meanings. An example is another Barnsdale, in Rutland, which was spelled as *Beonardeshull* in a legal record of 1202. *Beornheard* was an Old English personal name, and the original meaning resolves as 'Beornheard's hill'. A similar derivation for Yorkshire's Barnsdale is speculation, but it fits the facts on the ground.

The geography of early Robin Hood stories is usually general. Anthony Pollard points out that by the time the tales were first committed to writing they were 'stories with a northern setting compiled for and circulated predominantly among southern English audiences. As such, they could draw upon a long-standing stereotype of the "north".'[45] Pollard and others suspect a northern genesis for some elements that were woven into the stories, but note that 'there is frustratingly little evidence for their circulation in northern England itself'.[46] The *Gest*'s accurate knowledge of what turns out to be a localised neighbourhood is thus a puzzle. At whom were these local references aimed?

Another way to ask why precise reference points were provided for a readership that did not need them is to turn the question around. Mention of Robin Hood in *Piers Plowman*, a later fourteenth-century allegorical dream, or of the roadside stone in 1422 remind us that Robin had long been a legendary figure. If the *Gest* was designed to be recited in instalments,[47] what if someone in the vicinity adapted a selection of tales that were already widely known for a local audience? If such a collection was then printed and began to circulate it would appear that the stories had originated around Barnsdale.

Something along those lines would account for the incongruity of real places in an unreal setting, but in doing so it introduces another problem. We have no date for the manuscript behind the printed versions that appear from the 1490s, but it seems unlikely that a large text on such a popular subject could have existed for any length of time without attracting notice.[48] It follows that the *Gest* is likely to have been set down later rather than earlier in the fifteenth century. However, Barnsdale and Robin Hood were being bracketed together well before this. The Scottish poet and Augustinian canon Andrew of Wyntoun included a verse about 'Litil Iohun and Robert Hude' in his *Orygynale Chronicle* (1408–20), describing them as forest outlaws 'In Ingilwode and Bernnysdaile'.[49] Inglewood ('The Wood of the English') is readily identified as the forest of that name in Cumberland, where the Scottish crown held interests.[50] Scholars ponder the respective claims of the Yorkshire and Rutland Barnsdales. Whichever it was, the Stone of Robin Hood shows that early in the fifteenth century people between Doncaster and Pontefract believed it to be Yorkshire.

It was not just locals who made this association. A lawyer in the Court of Common Pleas used the line 'Robin Hode in Barnsdale Stode' in 1429 to characterise his opponent's pleading as nonsense.[51] The way in which he said this indicates his expectation that everyone knew the line.[52] Why that should be invites yet another discussion, but what it means here is that by the early fifteenth century 'Robin Hood' as a prompt in a word association game would elicit 'Yorkshire'.

We saw at the start that by 1350 'Robehod' was a generic name for a life-threatening wrongdoer. We have also seen that the Street, here joined by other trunk roads,[53] was a kind of elevated pass that made it ideal terrain for banditry. This gives an underlay for 'Robin Hood in Barnsdale stood', and does so for two reasons. One is that highway robbery in fourteenth- and fifteenth-century Yorkshire was endemic; legal records reveal 'a rash of incidents' in the decade 1462–72, including two holdups in the vicinity of Barnsdale and Wentbridge.[54] In the 1480s Wentbridge had at least one inn that was capable of providing accommodation for long-distance travellers;[55] the name would thus have been known to some from far afield. Here, then, is material for someone who later in the fifteenth century reworked already circulating tales about a criminal who was good, mapped them onto a well-known and longer-standing tradition, and projected a limitless greenwood onto an agricultural area.[56] The second explains why Barnsdale was familiar to southern audiences even if they had no idea what it was. Its position astride the road between Doncaster and Pontefract made it a place of meeting and mustering at times of stress and disorder, and thus a name that everyone knew. It is mentioned repeatedly, for instance, in state correspondence during the Pilgrimage of Grace (1536) and in subsequent confessions and examinations.[57]

The Went crossing and Barnsdale also formed a ceremonial threshold. In 1478, for instance, this was where Yorkshire's leaders turned out to meet Edward IV as he arrived on his first northern progress.[58] In

Robin Hood's topsy-turvy world where crooks administer justice, his meetings with travellers mirror the greetings of kings and magnates. Barnsdale Bar in the sixteenth century emerges as a place of popular symbolism, akin to today's Watford Gap, where one England ends and another begins.

And there we might end, and probably should, but for things left over. Spellings, rhymes and other textual indications put composition of the *Gest* in the north, and very likely in Yorkshire. Reading, writing and the Church went together – whence, 'clerical'. The majority of known fifteenth-century English writers were either ordained or members of religious communities. Whoever put the *Gest* on the page is thus likely either to have had an ecclesiastical background or education at the hands of churchmen. Institutions nearby could have provided that, like the house of Cistercian nuns at Hampole, or the Cluniac houses at Monk Bretton and Pontefract. Pontefract was also home to a Benedictine community and college of priests, while the chaplaincy at Pontefract Castle was institutionally connected with the parish church of Campsall.

What, indeed, of Campsall, in which much of Barnsdale lies and where the church, like Robin Hood's chapel, is dedicated to Mary Magdalene?[59] St Mary's is a large, intricate building that reflects hefty investment by two Norman elite families in the twelfth century and successive expansion. The original building resembled – and perhaps was intended as – a collegiate church, with side altars suggesting a team of several priests rather than a single incumbent. This would fit the picture of wealthy lords seeking commemoration in the long term.[60] There were also chantry priests in the area – chaplains who sang masses for the repose of the souls of individuals who left endowments for the purpose. They were supernumerary to the parish clergy (which at Campsall in the sixteenth century consisted of a vicar and a deacon[61]) and thus available for other duties – like teaching. A first-floor chamber above the vaulted west bay of the south aisle is just the kind of space that could have been used for the purpose. It dates from the end of the thirteenth century, when the emergence from the village of someone with the stature of the theologian Richard of Campsall (c. 1280–c. 1350) suggests a well-established tradition of schooling.[62] By the fifteenth century the villagers themselves had formed a fraternity – part social club, part cooperative chantry – hiring their own priest 'to pray for the parishioners living and the souls departed'.[63] The lord in the stone-built fifteenth-century manor house across the road had his own chapel, and presumably a chaplain. Villagers in Fenwick and Moss, townships out in the marshes between the Aire and Don, worshipped in a parochial chapel they called St Mary-in-the-Fields.[64] Just west of the Street there was another chapel, recently founded for the ease of villagers from Wrangbrook and Menscroppe 'in winter and foul weather'.[65]

Of course, for a rich literary culture and an audience we think first of York – which was home to the self-interested abbot who was success-fully scammed by Robin, and an early centre of printing.[66] Yet again 'the Saylis' draws us back to some more local connection. 'The Saylis', we recall, is an area in the township of Kirk Smeaton where Robin and Little John 'walk up' near the start of the *Gest*. While travellers moving through the area would be familiar with Wentbridge and the Street, it is hard to see how they would come to know the name of a field. 'The Saylis' implies very local knowledge.

Several places closer to hand might provide the kind of context we seek. Doncaster is just seven miles down the Street. In the 1540s over 2,000 people lived there, supported not only by parish clergy but by six chantry priests who were attached to the large parish church, and a chapel in the marketplace which was home to the Guild of St Thomas.[67] Or what of Rotherham, a large market town which in the sixteenth century was known for outstanding 'smiths for all cutting tools', and for the habit of its citizens to burn coal rather than wood because coal was locally plentiful 'and sold good cheap'.[68] John Leland was much taken by the stately parish church, wherein six priests prayed for souls and ministered to townsfolk, and even more so by the College of Jesus that was 'sumptuously builded of brick'.[69]

Jesus College was begun in 1482/3 by Thomas Rotherham (1423–1500), who was appointed chancellor of the University of Cambridge (1469–71 and 1483–5), archbishop of York (1480–1500) and chancellor to Edward IV (1475–83). Thomas Rotherham founded the college on the site of his birthplace in thanks for the life-changing moment in his boy-hood when he and his friends 'stood there without letters', and 'a man learned in grammar' (very likely a chantry priest) arrived 'from whom as from a spring, we drew our first instruction'.[70] Schooling was accordingly a leading function of the college, which took in children 'of the poor sort' who were 'to be brought up in knowledge of grammar, song, and writing' which included both grammar and poetry.[71] Rotherham went out of his way to provide education in writing and 'reckoning' (business accountancy) for youngsters who did not wish to enter the Church. He endowed the college with a library of over a hundred printed books and manuscripts.[72] Of course, we cannot connect the *Gest* author with the fellowship of Jesus College, Rotherham, or anywhere else. All we can say is that Rotherham offered the kind of milieu, in the right area and at the right time, from which a secular author could have emerged. To judge from regulations and finds made at other late medieval colleges their members enjoyed pursuits that included gambling and popular music, and wide open to *Gest* culture.[73]

Jesus College, Rotherham, was disbanded in 1547–8.[74] The building was successively reused as a mansion, a malting house and an inn, and eventually swept away by the laying out of a new street.[75] However, there are other buildings in the area that reflect the degree to which it was admired. One of them is the mansion of a former monastic grange at Thurnscoe, a village in the Dearne valley between Doncaster and Barnsley once renowned for its cheese, later for its collieries, later still for the consequences of their closure, and about which Nikolaus Pevsner could find only twenty-six words to write.[76] Pevsner missed the grange, which is a pity for he might have spotted that its description in the statutory list as 'probably late C16 with C17 additions' underestimates its age:[77] Low Grange's earliest fabric of diapered brickwork very likely dates from before the Reformation, belonging to the world of Thomas Rotherham when bricks and printed books were signs of things to come.[78]

The rymes of Robin Hood were expressions of nostalgia for a happier place in an age that never existed: a Maytime of birdsong, fellowship and topsy-turvy order whereby justice is administered by outlaws led by a 'good thief'. The rymes immortalised people who probably never lived, yet will be remembered long after the memories of those who did live have faded. They were written down in an age of veiled meanings. Rotherham's College consisted of a provost, three fellows and six choristers, making ten, so that as Thomas Rotherham wrote in his will, 'where I have offended God in his ten commandments, these ten should pray for me'.[79]

There were others further off who would have appreciated that: among them were the warden and fellows of Merton College, Oxford, whose far-flung assets included lands in Northumberland that called for periodic tours of inspection. College accounts in 1464 give a glimpse of college members resting at Wentbridge on both outward and return journeys.[80] The inn beside the bridge, ancestor to Barnsdale Bar's Little Chef, would be just the place to swap yarns with locals and pick up scraps of detail; Merton College, Oxford, would be good place in which to write them down.

MUNGO, SHODDY AND ENEMIES WITHIN

—————

S econd Lieutenant J. B. Priestley was twenty-five when he was demobilised in March 1919. Blinking as he stepped back into 'civilian daylight', he returned to his home in the city of Bradford.[1] A place awaited him at Cambridge; in the months before taking it up he thought of a walking tour in the Yorkshire Dales and wrote a weekly column for the *Yorkshire Observer*.

The *Observer* had been started eighty-five years before as the *Bradford Observer*, a Liberal-inclined weekly whose editor promptly used it to attack the fiery Leeds-born campaigner Richard Oastler (1789–1861).[2] Oastler's causes included the abolition of slavery, Roman Catholic emancipation, resistance to the Poor Law Amendment Act 1834, and, above all, factory reform. In 1830 he had written to the *Leeds Mercury* denouncing worsted mill-owners in and around Bradford as perpetrators of child slavery. By doing so he put the cause of industrial regulation on a national footing, and opened a 'war against factory wrongs' that continued for seventeen years.[3]

Two years later Oastler called attention to the contrast between current factory practice and the kind of family-based manufacture which had been practised during his childhood:

> it was the custom for the children at that time, to mix learning their trades with other instruction and with amusement, and they learned their trades or their occupations, not by being put into places, to stop there from morning to night, but by having a little work to do, and then some time for instruction, and they were generally under the immediate care of their parents; the villages about Leeds and Huddersfield were occupied by respectable little clothiers, who could manufacture a piece of cloth or two in the week, or three or four or five pieces, and always had their family at home: and they could at that time make a good profit by what they sold;

there were filial affection and parental feeling, and not over-labour . . .

But now, he said, 'that race of manufacturers has been almost completely destroyed':

> On one occasion I was very singularly placed; I was in the company of
> a West India slave master and three Bradford spinners; they brought
> the two systems into fair comparison, and the spinners were obliged
> to be silent when the slave-owner said, 'Well, I have always thought
> myself disgraced by being the owner of black slaves, but we never, in
> the West Indies thought it was possible for any human being to be so
> cruel as to require a child of nine years old to work 12½ hours a day;
> and that, you acknowledge, is your regular practice.
>
> I have seen little boys and girls of 10 years old, one I have in my
> eye particularly now, whose forehead has been cut open by the thong;
> whose cheeks and lips have been laid open, and whose back has been
> almost covered with black stripes; and the only crime that that little
> boy, who was 10 years and 3 months old, had committed, was that he
> retched three cardings, which are three pieces of woollen yarn, about
> three inches long. The same boy told me that he had been frequently
> knocked down with the billy-roller, and that on one occasion, he had
> been hung up by a rope round the body . . .[4]

Oastler answered the *Observer* in a pamphlet preceded by this announcement:

> IN THE PRESS, AND SPEEDILY WILL BE PUBLISHED,
> *PRICE FOURPENCE*, A LETTER To those Sleek, Pious, Holy
> and Devout Dissenters, Messrs. Get-all, Keep-all, Grasp-all, Scrape-
> all, Whip-all, Gull-all, Cheat-all, Cant-all, Work-all, Sneak-all,
> Lie-well, Swear-well, and Company – **The Shareholders in the
> Bradford Observer**, in answer to their Attack on R. Oastler, in that
> Paper of July 17th, 1834.
>   By RICHARD OASTLER[5]

Free-market Liberalism obstructed legislation on the working hours of women and children for another thirteen years.[6] When Oastler died in 1861, the *Observer* implied that he himself had been partly to blame for the delay, on the grounds that the 'unmeasured' manner of his campaigning had alienated many who might otherwise have supported 'moderate restrictions'. By this date, the *Observer* had accepted the case for legislative intervention, although in its view Oastler had regarded

mill-owners generally as 'men of tyrannical dispositions' and as the
'sworn enemies' of their employees.[7]

It is interesting to compare these qualifications with accounts of
Oastler's funeral. The family's wish was for a private ceremony. When
the cortège set forth from Harrogate at about nine in the morning on
30 August it consisted of just a horse-drawn hearse and two coaches
carrying nine relatives, headed for the church at Kirkstall, thirteen miles
away, where the funeral was due to begin at noon. When the mourners
reached Headingley around 11 a.m. hundreds of 'steadfast and faithful'
personal and political friends were waiting. Among the friends were
representatives of practically every Short Time committee from Lancashire and Yorkshire – the men and women who had campaigned with
Oastler a quarter of a century before, when the Ten Hours movement
was in its earliest and most difficult stages. With them were MPs,
lawyers, churchmen, and mill-owners. They formed up in procession
and walked on with the cortège between growing knots and crowds of
onlookers. As they neared the church, hundreds of mill-hands, factory
workers and children stood at the churchyard entrance. Twelve factory
workers, six from Lancashire, six from Yorkshire, stepped forward and
lifted the coffin.

> When the hearse had been emptied of the coffin, and the body
> borne by Labour's sons into the church, through the opened ranks
> those who had preceded it from Headingley, the utmost anxiety was
> manifested by the assembled Factory operatives – it was just 'dinner
> hour' – to be permitted to enter the churchyard. For a period, it was
> attempted by the police in attendance to keep the gates shut . . . Time,
> however, to workers was precious; and at length their importunity
> became so urgent that the gates were thrown open, and the anxious
> crowd entered the churchyard. Those of them who could spare the
> time took up places in the Church gallery; but the great mass, having
> to obey the call of the Factory-bell, had content themselves with a
> running-look at the grave in which the body of their 'Old King' was,
> in a very few moments, to be deposited, and then depart to labour . . .[8]

The funeral service was led by 'Parson Bull', a former vicar of St James,
Bradford, and longstanding ally who had been effective chaplain to the
Ten Hours movement. In following days Bull preached more funeral
sermons, in Bradford and Huddersfield. One journalist accounted the
occasion 'thrilling', and 'so free from the make-believe of sorrow, as to
be fully in accordance with the simple, upright, but withal grand nature

of the man'. Oastler was buried beside his beloved wife, Mary, and their two children, who had died in infancy.

Two small incidents at the close might have been in a Priestley novel. After years of struggle and a term of imprisonment for debt that resulted from misfortune, Oastler had retired to Conway in north Wales. Here, he grew flowers. A rose, some annuals and a sprig of myrtle were forwarded by an 'old servant' who wrote that 'the very flowers in the garden drooped, now that the master is gone'. The undertakers put the blooms under Oastler's coffin plate, as hidden emblems of publicly recognised virtues.

Another episode recalled early days of the Ten Hours campaign, when Oastler's coat had been ripped during an altercation with Lord Macaulay at a political meeting in Leeds. Afterwards, a factory child ran up offering a pin to join the sundered parts. Oastler had lifted the pin high and told the crowd that with it he would fix the Ten Hours Bill to the statute book of England. As he went into the grave, the pin rested on his heart.

Neither of these happenings was mentioned by the *Bradford Observer*, which seven years later, under both political and commercial challenge from the new *Bradford Daily Telegraph*, became a daily. By the end of the nineteenth century it circulated throughout the region and was read further afield for its authoritative coverage of matters to do with wool.[9] It was renamed the *Yorkshire Daily Observer* in 1901, and became the *Yorkshire Observer* in 1909 following its acquisition by the West Riding wool magnate and Liberal politician James Redman Hill (1849–1936).[10]

Hill was very well known in Bradford, where he had served as a magistrate, as a councillor (since 1898), as Lord Mayor (1908–9), and most recently as the Liberal MP for Bradford Central (1916–18). Later, he became a director of Barclays Bank, and leader of the syndicate that in 1918 acquired Salt's Mill and its business for a price not far short of £2m.

Local historians bewailed Bradford's obsession with material achievement at the expense of literature or science.[11] John Ruskin might have had Hill in mind when he said that Bradford's paramount devotion was to 'the great Goddess of "getting on"'.[12] During Hill's three years as an MP, the sum of his activity on the floor of the House was two questions to the President of the Board of Trade about wool supply.[13] Nonetheless, Hill typified another feature of northern capitalism: by combining commercial and civic life with presidencies of trade organisations, public service, philanthropy and religious example, he epitomised the idea of what a thriving northern industrialist should be. Thus, he chaired the city's Tramways Committee, led fund-raising for Bradford's Infirmary and Children's Hospital and gave a park and allotment gardens to his home district. Like many members of Bradford's millocracy he was a dissenter; he became president of the Bradford Free Church Council and for many years taught in his local Congregational Sunday school. The newsworthiness of his death in January 1936 was somewhat outdone by the passing of Rudyard Kipling on the same day and George V three days later. Nonetheless, the *Yorkshire Post* printed its eulogy, headlined 'A pioneer of modern Bradford', alongside the obituary of 'Kipling, the poet of Empire'.[14]

Sir James used the gains of industry to enter the life of a country gentleman. In 1918 he and his wife, Alice, acquired Hexton Manor, a mansion on the slopes of the Chilterns in an estate of nearly four square miles wherein were avenues of lime and beech, clipped box hedges, sweeping drives, lakes and pleasure grounds. Six years later, their elder son, Arthur, moved to Denton Hall, a country house in parkland eleven miles north of Bradford, where he cherished a herd of British Friesian cattle, rejoiced in his reputation as the best-dressed man in Bradford trade, and entertained the future George VI to shoot on his estate. The transition from an urban mill to a rural manor was another aspect of the role to be played by those who 'got on'.

There is a sense of whiplash about the speed of James Hill's passage from poverty to the royal fringe. He was born out of wedlock to Margaret Redman, a weaver, in the village of Harden on Bradford's edge. His father was William Hill, a 23-year-old clog-maker. Margaret and William were probably already living together and in due course they married; three more children followed. By the age of ten James was working as a greengrocer's boy, visiting Bradford's wholesale market twice weekly with a donkey and cart, to select produce and take it back to Harden. The self-reliance this taught may well have influenced what followed. From twelve he worked in one of Harden's mills as a trainee weaver, then as an apprentice wool sorter, where he found himself acting as 'foreman and salesman, as well as sweeper-up and firelighter'. At eighteen, apprenticeship completed, he had saved £50 and used the money to become a wool-buyer. Two years on, he joined John Reddihough, a firm of wool merchants, top-makers and wool-combers, where he stayed for sixteen years and became a partner. In 1891 he founded his own firm, James Hill and Sons, which imported wool and later added wool-combing and top-making to its interests. By the time young Jack Priestley joined his newspaper's payroll, Hill had been created a baronet and his business was world-famous.[15]

James Hill's childhood and earlier career were roughly paralleled by events in Bradford's development. We have already met examples of conditions in the 1830s and 1840s – Bowling Hell, Benjamin Myers on his knees hauling coal tubs at the age of eleven, polluted water, and William Scoresby's breakdown in the course of a struggle for improvements at which free-market mill-owners sneered. But, in due course, things did improve. The formation of a borough corporation in 1847 was followed by the acquisition of a water company and the introduction of building regulations. In the 1860s city centre streets were widened. Sewers and drains were laid. In the 1870s came a free library, art gallery and museum, the beginnings of slum clearance, and a town hall recalling the world of thirteenth-century Gothic. By now there was greater social differentiation, an expansion of professions – architects, teachers, lawyers, engineers – and growing attention to public health. In 1889 the corporation provided a municipal electricity supply.[16]

At the end of the Second World War, Priestley looked back at someone rather like himself, in a fictional place he called Bruddersford that was rather like Bradford, in the 'great gold Maytime' before 1914. In his novel *Bright Day* (1946) Bruddersford was remembered as a place of fellowship, haloed with a glow that had since faded. During the 1930s, for instance, Christmas had become a commercial racket that began in mid-autumn. Back then:

> Christmas arrived at the proper time, late on the twenty-fourth of December, but once it did arrive then it really was Christmas – and often with snow too. Brass bands played and choirs sang in the streets; you went not to one friend's house but to a dozen; acres of rich pound cake and mince-pies were washed down by cataracts of old beer and port, whisky and rum; the air was fragant and thick with cigar smoke, as if the very mill chimneys had taken to puffing them; whole warehouses of presents were exchanged; every interior looked like a vast Flemish still-life of turkeys, geese, hams, puddings, candied fruit, dark purple bottles, figs, dates, chocolates, holly, and coloured or gilded paper hats; it was Cockaigne and 'the lost traveller's dream under the hill'; and there has been nothing like it since and perhaps there will never be anything like it again.[17]

47. Alhambra Theatre, Bradford, 1949

Bright Day evokes a 'rich, democratic and self-sufficient cultural' world of 'Free Libraries, Playgoers' Societies, Hallé Orchestra concerts in the Gladstone Hall, pantomime at the Theatre Royal and the "brilliant Indian Summer of a popular art" at the Imperial Music Hall' (Fig. 47). Bruddersford was no oop-north backwater but a cosmopolitan city whence even 'modest employees' travelled the globe to trade in wool, where 'a Londoner was a stranger sight than a German'.[18] Bradford's Little Germany is one of the most unexpected and special places in Europe: a family of hillside streets in a neighbourhood a fifth of a mile across, lined with lofty warehouses embellished with ironwork, pink granite and carved motifs.[19]

In contrast, 'the town' recalled week by week by young Priestley in 1919[20] was a place of ceaseless noise, 'choked with fumes', uncomfortable in summer because it was cram-full of 'pushing, sweating' people, 'blackened' by 'mechanical activity' and wreathed in ennui. No wonder, then, that alongside Bradford's standing as Yorkshire's shock city of the industrial revolution was its reputation as a place to get out of (Fig. 48).[21] Escape is the intimation of the cover of the first edition of *The Good Companions* (1929), which shows a thicket of smoking chimneys rising from a sooty smudge whence runs a white ribbon of oh-so-inviting open road to somewhere else. At the book's end, Jess Oldroyd goes to Canada.

Something more complicated lay behind the binary. West Riding towns were usually situated in valleys, or in Bradford's case in a bowl; between the valleys lay uplands, 'fresh from the mint of heaven', where birds sang, the air was 'pure, almost intoxicating', and you could look into 'purple distances'. Priestley repeatedly celebrated the 'high places'

48. Bradford at work

intercalated between West Riding conurbations. In *Bright Day*, for instance, nature was presented not as Bruddersford's opposite but as its companion:

> Lost in its smoky valley among the Pennine hills, bristling with tall
> mill chimneys, with its face of blackened stone, Bruddersford is generally held to be an ugly city; and so I suppose it is; but it always
> seemed to me to have the kind of ugliness that could not only be
> tolerated but often enjoyed; it was grim but not mean. And the moors
> were always there, and the horizon never without its promise. No
> Bruddersford man could be exiled from the uplands and blue air; he
> always had one foot on the heather; he only had to pay his tuppence
> on the tram and then climb for half an hour to hear the larks and
> curlews, to feel the old rocks warming in the sun, to see the harebells
> trembling in the shade.

Trams in Priestley's writings were sometimes brokers between industrial towns and the adjoining high places, akin to the huntsmen and hermits who mediated between civilisation and wilderness in medieval romances. In the opening of *The Good Companions* (1929), swooping down on Bruddersford as if through Google Earth, introducing men and women 'who are afraid of nothing but mysterious codes of etiquette and any show of feeling', Priestley paused to contemplate the valley slopes:

> If it were night, you would notice strange constellations low down in the sky and little golden beetles climbing up to them. These would be streetlights and lighted tramcars on the hills, for here such things are little outposts in No Man's Land and altogether more adventurous and romantic than ordinary streetlights and tramcars. It is not night, however, but a late September afternoon. Some of its sunlight lights up the nearest of the towns, most of it jammed into a narrow valley running up to the moors.[22]

The moors themselves, of course, had been busy industrial sites, quarried for millstones; poked, pocked, scarred and hushed for lead; but by the time Priestley took Sunday walks on Baildon Moor or hiked up Wharfedale that was over. The moors:

> Are a stage
> For the performance of heaven.
> Any audience is incidental.[23]

Most of Priestley's articles for the *Observer* were written under the pen name 'Peter of Pomfret', an early thirteenth-century hermit credited with gifts of prophesy.[24] Why Priestley chose this alias (or indeed why medieval hermits should have resided near the entrances of so many West Riding dales[25]) is a mystery, but it is not difficult to think of possibilities, such as the mask provided by an unfamiliar name, or because he enjoyed the way it sounds. Whatever it was, Pomfret – that is, Pontefract (or, as locals know it, 'Ponty') – maps straight into Yorkshire industry as the town that historically controlled southern access to the confluence of the rivers Aire and Calder. We have already seen how from about 1700 it was the adaptation and augmentation of these rivers that linked Leeds, Wakefield and adjoining towns with the sea, and so facilitated the bulk shipment of manufactured goods to other parts of Britain and the near continent.[26] Extensions through the Calder and Hebble Navigation and the Leeds–Liverpool Canal linked the industrial West Riding with the world (Fig. 49).

Ted Hughes likened Calderdale to one long street of industrial towns.[27] Most of the addresses in the West Riding 'heavy woollen' district either stand along this street or, like Cleckheaton and Heckmondwyke,[28] are up one of its side roads. Liversedge, Mirfield, Batley and Huddersfield,

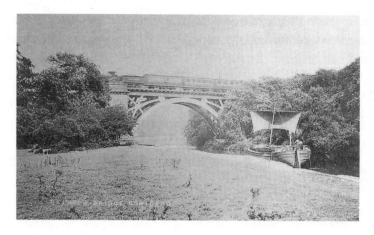

49. *Rainbow Bridge with Humber keel, Conisbrough*

for instance, all lie on Calder tributaries that drove their mills before coal-fired steam engines took over.

At the start of *English Journey* (1934) Priestley compared sparkling white art deco factories with the image of 'a grim blackened rectangle with a chimney at one corner' that had been fixed by his Bradford boyhood.[29] The concept of such a 'proper factory' went back to the late eighteenth century, when a start was made to bring tasks hitherto performed separately under one roof. Attempts to mechanise stages of textile production had been tried for decades, but it was pioneers like Benjamin Gott (1762–1840) who brought them together. In 1792 Gott introduced a system on a site in Leeds (at Bean Ing Mill) in which the entire sequence of scribbling, carding, fulling, spinning, dyeing and finishing was integrated. In result, forty years later, virtually everything that mattered to people was changed: attitudes to time, where you lived, family life, ecology, public health, social relations – and Richard Oastler and Michael Sadler were grappling with the consequences. Even pathogens changed. By 1850 Bradford was importing wool from Iran, Russia and South Africa, and with it, sometimes, came spores of anthrax.

The ubiquity of the factory format, and generalising labels like 'heavy woollen', give an impression that the same things were going on all over the West Riding. In fact, as Asa Briggs emphasised, industrialisation did not homogenise nineteenth-century towns so much as tell between them.[30] Different places developed individual cultures and traditions. Contrasts and rivalries emerged among their elites, the imagery and messages on their town halls (Plate 23), the architecture of their railway stations, housing (Fig. 50) – divergences extended even to the vagrant plants that grew on the margins of mill-yards and towpaths. Communities of urban weeds evolved variously from town to town, according to their land-use histories, or the character of their parks and allotments. An ecologist led blindfold onto waste ground in Sheffield could tell it apart from Bradford.[31]

Similar intricacies are found among the industries themselves, many of which became co-dependent. Leeds flax spinners importing Baltic flax through Hull and the Aire and Calder navigation fostered the design and manufacture of textile machinery that in turn yielded linens of differing fineness for assorted markets. High-volume buying, selling, customer contact and accountancy were all reliant on paper record-keeping, so creating demand for stationery, thus for printers, and so for printing machines. The need for up-to-date information in always-changing commercial environments similarly gave opportunity for publishers of newspapers and directories. Among the mills producing woollen and

50. Pioneering social housing: Quarry Hill Flats, Leeds (begun 1934) was a modernist-inspired complex to replace slum housing. The scheme included many technical innovations, aimed to house 3,000 people and to include communal and sports facilities. However, work was interrupted by the war and was never fully seen through. The flats became physically isolated and were demolished in 1978.

worsted yarns were others that specialised in complementary processes and equipment. These included makers of blankets and carpets; dyes; chemicals; leather and steel wire for scribbling and carding; belts, pulleys and gears for line-shafting. Lower down the dales, where the rivers emerged from the Pennines and began to meander towards their rendezvous with the Humber, clusters of specialised industries made things like bottles and jars, mustard or porcelain. Hallamshire and towns in the triangle formed by Barnsley, Rotherham and Doncaster supplied the world with cutlery, scissors, sickles and all manner of cutting tools. In the middle of it in the 1940s was the young Ted Hughes, avidly reading the comics in his father's newsagent's shop in Mexborough, fishing local ponds, storing up experiences that would later be released in a lifetime of writing.[32]

Aback of it all, of course, was coal: the abundant, on-the-spot fossil plant matter that had been laid down in deltaic swamps all those hundreds of millions of years ago,[33] and which from the commercial development of the steam engine in the eighteenth century became the fuel that powered the making and carrying of almost everything. At the end of the nineteenth century the West Riding contained over 450 working coal mines (Plate 25).

Batley, Morley, Dewsbury and Ossett specialised in rag collection and sorting for the production of shoddy and mungo. Shoddy is recycled wool, recovered from textile waste by a grinding process that was introduced in 1813 and respun as yarn. Machinery devised in 1835 to mince hard rags, old clothes and tailors' offcuts yielded fibrous material for another fabric called mungo. Mungo and shoddy could be blended with new wool, while a cotton warp and a mungo/shoddy weft could be combined as union cloth. Such a hybrid fabric was cheaper and coarser than textile woven from pure fibres, and thus ideal for mass-produced garments like uniforms and greatcoats. For many decades, a large proportion of the world's police, armed forces and marching bands was clothed from Yorkshire (Fig. 51).

At the far extreme from this bristly material was silk – a natural fibre, and a word with a worldwide root – and derivative fabrics like velvet, chiffon and taffeta. Jack Priestley's mother, Emma, came from a silken world.[34] Before her marriage to Jonathan Priestley in 1891 she was Emma Hoult, a silk-picker who lived with her family in Silk Street, in the Bradford suburb of Manningham. Her sisters Eliza, Sarah and Anne were silk workers (a silk-weaver, blush-winder and silk-picker, respectively), while her father, Thomas, worked as a stoker at the silk

51. Local band, Bradford, c. 1912

mill where he played his part in running the steam boilers that drove the looms. This was a sizeable task (the boilers consumed about 166 tons of coal each working day), and to be on the safe side the mill had its own coal mine, at Featherstone near Pontefract. Smoke from the boiler fires was conducted to the sky by a 250-foot chimney that towered above the city and reminded its citizens of an Italian campanile (Fig. 52). The mill's floor area covers nearly half a square mile, and in 1893, the year of Jack Priestley's birth, 5,000 people were working on it. No wonder, then, that Manningham mill has been described as one of the most theatrical expressions of industrial prowess anywhere in Victorian England.[35]

Manningham's master was Samuel Cunliffe-Lister (1815–1906), whose life embraced the Battle of Waterloo and the Wright brothers. Lister's wealth at death was only £100,000 or so more than that of James Hill, but their origins were very different. We recall that Hill was born with nothing and was already working at the age of ten. Lister's father was a wealthy manufacturer who sent his son to a private school, supported his early career as a merchant, and in 1838 built a worsted mill for him and his brother John to run. When John inherited the family estate, Samuel carried on, initially in partnership with others and from 1864 on his own. The great silk mill was built in 1871–3, following the loss of its predecessor in a fire.

Samuel Lister was interested in processes. Visits to the United States in his merchant days taught him about American business methods,

52. Lister's mill, Manningham, 1951

and much of his long career was devoted to invention. He patented
over 150 creations, sometimes at a torrential rate. In 1855 he began to
think about the possibility of using the silk waste that is left when the
thread has been unravelled from a cocoon. Such waste was cheap – at
that point a halfpenny a pound – and Lister worked on a machine to
convert it to usable material. This took time, and lack of interest in the
silk industry brought him close to bankruptcy. However, in 1864 his
silk recovery process began to pay. Lister bought silk waste from India,
Italy, China and Japan, shipped it to Bradford, and used his machines
to transform it to poplins, velvets, plushes and carpets. When Jonathan
Priestley proposed to Emma Hoult, Lister's annual profit had reached a
quarter of a million pounds.

Lister used part of his wealth to purchase the estates of Middleham
Castle, Jervaulx, and Swinton Park near Masham. All three lay on the
margin of the Vale of Mowbray, and since they adjoined he became
master of a single land extending from north of Ripon to the mouth

of Wensleydale. Thus did Lister step into the shoes of families like the Conyers and Nevilles who had been seated thereabouts 400 years before. And if the name Cunliffe-Lister sounds familiar, it is because we have already met it.[36] Philip Cunliffe-Lister, patron of the memorial window with a moorland Pentecost, the man who in 1936 equipped Britain with Spitfires and radar, was the husband of Samuel Lister's granddaughter.[37]

Late Victorian Bradford was led by Nonconformist industrialists who made few concessions to their workforces. There were exceptions, like Titus Salt (1803–76) who moved chosen workers out of Bradford onto a greenfield site near Shipley, where he had built them a village with baths, an institute, churches and hospital, but the prevailing climate was one of economic liberalism. (Forty years back, the argument against Ten Hours legislation had been that if Bradford factory masters were unable to employ child labour, they would be out-competed by other employers who could.) Down to the 1880s West Riding textile concerns were not much unionised, and the separate specialised processes that went on inside them enabled employers to exploit differences and prevent work-ers from uniting in opposition. This was helpful when they wished to

53. Undercliffe cemetery, resting place of many nineteenth-century Bradford elite

lower their costs, as happened on 9 December 1890 when Samuel Lister notified 1,100 workers in his velvet department that their wages were to be reduced in response to the increase of duty on goods entering the United States.

The velvet workers went on strike on 16 December. Few of them were union members, but the West Riding Association of Weavers organised public meetings and a strike fund. A standing committee was elected to negotiate. Collections were made. Workers travelled out to canvass support. There were marches through the city, to draw attention to their cause and to meetings in the auditorium of the Star Music Hall. A reporter for the *Manchester Guardian* watched one of them:

> This afternoon the workpeople who are out on strike at Manningham Mills made a 'demonstration' in the form of a procession, to a band of music, in the streets of Bradford. One striking feature of the procession was the very large proportion of well-dressed women which it included. On a placard carried in the procession it was stated that 1,200 persons were out of employment in consequence of an intended reduction of wages ranging from 12½ to 30 per cent. The firm at Manningham Mills allege that the reduced wages offered are quite equal to those paid by all other firms in the district, and that they are compelled to make the proposed reduction in order to meet the adverse effect on their business created by the new American tariff.[38]

On 19 January, 1,097 out of 1,100 plush weavers rejected a marginal concession. The strike committee challenged Lister's claim that Manningham's rates were no less than those for equivalent work elsewhere. Early in February its chairman reported that in Rochdale 'he had found as many looms engaged on silk and plush goods as at Manningham Mills, and that the operatives were better paid, and more regularly employed, under conditions of life which were better than those which had existed at Manningham Mills for the past fifteen years'.[39] The dispute was mostly mannerly, but tensions grew. On one occasion the firm's directors were mobbed. By March, Lister's were bringing in strike-breakers, described by the strikers as 'few and incompetent', and by the end of March the entire workforce was out. Lister remained intransigent. The city Watch Committee began to impose restrictions on assembly. On 13 April the strike committee determined to hold a public meeting in an open space beside the Town Hall, the offer of another and more distant site by the mayor and Watch Committee having been declined. In the late afternoon the gathering shrank, then regrew through the evening. As it did, some members began to attack the police, who called for reinforcements and horses. Bricks, stones and knives were thrown at the police, several of whom were seriously injured, and one of the horses slain. Towards nine o'clock the mayor sent a messenger to Bradford Moor Barracks, calling for assistance. The Durham Light Infantry sent 160 men, each armed with a Martini rifle and forty rounds of ammunition. Little by little the soldiers cleared the square. Skirmishes continued to the end of proceedings that were described next day as having 'no equal in the modern history of Bradford'.[40] A peaceful demonstration was held the following week, but the strikers were worn out. By 27 April their action was practically done, ended by hunger, hardship, threats of eviction and the formidable resources of the company.

Samuel Lister and his directors no doubt perceived the strike as a self-contained episode, but there were consequences that neither they nor Bradford's Liberal elite could have foreseen. The defeat shook manual workers and brought trades union organisers together. Within a month, the Bradford Labour Union had been formed; within two years, the Independent Labour Party was in being. In 1906 Labour won the parliamentary seat of Bradford West. The Manningham strike led to fundamental political realignment, one eventual result of which was the nationalisation of Britain's coal industry (Figs. 54, 55).[41]

Did Margaret Thatcher remember any of this when, at about 5 p.m. on a July Wednesday in 1984, she walked into a room in the House of Commons to address Conservative backbench MPs? This was the occasion when – according to her notes – she described the leaders of the National Union of Mineworkers (along with politicians in some local authorities, and in contradistinction to the recently defeated junta in Argentina) as 'the enemy within'.[42] The great coal strike was then in its nineteenth week. South Yorkshire simmered. Police stations in three towns had been besieged; one town had been sealed off; a man had died (Fig. 56). Many historians conclude that the authorities themselves conspired to draw mass pickets into conflict with the police at Orgreave. As at Manningham, the action ended in collapse.

When the strike began, 56 of Britain's 173 working pits were in Yorkshire. In 2015, none. The only coal mine it is now possible to enter is a museum.[43]

The defeat changed Yorkshire as a landscape, as a community and as an idea (Fig. 57). It also changed Britain. And like the end of the Cold War with which it roughly coincided, we still have no sure idea what this means.

54. Wentworth Woodhouse, a great country house in parkland, rebuilt by the Marquess of Rockingham and inherited by the earls Fitzwilliam, photographed in 1926

55. Wentworth Woodhouse in 1947, following open-cast mining at the instigation of Emanuel Shinwell, Minister of Fuel and Power in Clement Attlee's government

56. Police discourage striking miners from travelling to picket, 1984

57. Colliery spoil heap, 2017. Such tips exist over much of western and southern Yorkshire, but now bear trees and grass, and are often used for recreational purposes. Many visitors assume them to be natural features.

12

ALL FOR YOUR DELIGHT

B etween the estuaries of the Tees and the Humber are soaring headlands and rocky foreshores, fishing villages astride deep inlets, remote
beaches, still-working ports, and boisterous resorts.

Behind Hull, there is Europe's fastest eroding shoreline, where truncated roads point to places long since taken, and elderly couples hold
out in bungalows that teeter on the edge of crumpling cliffs.

At the end of it all comes the remote, dynamic other-world of Spurn,
with its lighthouses, migrating birds and soughing marram grass.

Yorkshire people have strong feelings for their coast. In the 1890s or
1930s a day trip or works outing to Scarborough was something to be
dreamed about beforehand and savoured for months afterwards (Fig.
58). For many, nothing since has compared with the three-quarter-mile
ride on Scarborough's North Bay miniature steam railway, from Peas-
holme Park to Scalby Mills, followed by fish and chips at the Corner
Café. After 1945 when leisure was gregarious tens of thousands of
families holidayed each year at Billy Butlin's Filey camp, leaving soiled
cities by train and alighting at the camp's own station.[1] They also went
west. In the days when whole towns went on holiday together some of
Yorkshire's best bits of seaside were in Lancashire. Morecambe is still
known to many as Bradford-by-the-Sea.

In the 1970s cheap air travel to places with predictable weather did for
the English seaside holiday. And yet gregarious affection for Yorkshire's
coast has persisted through other means. Today, there are more than
forty caravan parks and mobile home villages between Scarborough and
Spurn, and it is probably no coincidence that Britain's largest manufacturer of static caravans – marketed as 'holiday homes and lodges' – is
on the edge of Hull. Many of the caravan parks are larger than the
places from which they take their names. Some have names of their
own, like Blue Dolphin or Crow's Nest, and in areas where parks abut
there are gatherings the size of towns. At weekends and during summer

58. Friends and neighbours together, 1939: knowledge of the individuals concerned suggests a Sunday School, Guild or Fellowship outing – another kind of family. David Pickard (far right) later trained as a navigator in Canada and was shot down and killed in July 1944

fortnights this is where many Yorkshire people go, and have done for years (Plate 26). The coast is also to be savoured by the day – an afternoon at Runswick with grandchildren, Withernsea's castellated entrance to a pier that is no longer there, high seas at Sandsend – or even by the minute: the ache of pleasure from that first glimpse of the North Sea as you come off the Wolds, or the memory of a first kiss.

It has long been so, although when my wife's Uncle Bernard was a youngster a family holiday on Yorkshire's coast was something he could only dream about. Robert Bernard Mitchell was born on a February Tuesday in 1898. Maskell Mitchell was his father. Maskell came from Darlington, and since the age of nineteen he had worked successively as a clerk, a booking clerk for the North Eastern Railway and now as a signalman. At the time of Bernard's birth, he was in a signal box on the approaches to Harrogate.

Since the sixteenth century Harrogate had grown famous for its spring waters charged with traces of iron, sulphur, and salt. Around 110,000 gallons a day issue from these springs, and word of their medicinal properties attracted visitors. Harrogate came to be known as the English Spa, and with the coming of railways it was a destination with which competing rail companies naturally wished to connect. By Maskell's day Harrogate was frequented by Europe's elite, attracted by the town's reputation 'for health and happiness' and by each other. His signals controlled their comings and goings along lines laid over forty years before by different concerns. One of them led onto the viaduct bearing the track built by the York and North Midland Railway across the valley of the Crimple Beck; another continued down the valley to Starbeck, where it met the line built to York by the East and West Yorkshire Junction Railway, and on to Ripon and Thirsk.

Between trains, Maskell Mitchell designed and tied fishing flies. He pored over the weekly *Fishing Gazette*, which declared itself 'Devoted to Angling, River, Lake and Sea Fishing, and Fish Culture'. Something else to read between trains was the *Yorkshire Weekly Post*, which, along with stories by current authors like Wilkie Collins and Rider Haggard, carried articles on science and agriculture, and an angling column by Thomas Evan Pritt. Pritt was a former banker from Lancashire, whose book on trout flies had shown how the flow and insect life of Yorkshire's becks and rivers differed from the gliding waters of southern chalk streams, and how a north English tradition of fly-fishing had evolved in result.[2] Pritt followed *Yorkshire Trout Flies* with *The Book of the Grayling*. He began:

> It has always seemed to me that the Grayling is entitled to a better place in the estimation of anglers than the one usually accorded him ... The trout is the angler's fish of spring and early summer, when every soft breeze is laden with a perfumed invitation to see Nature at her best; the Grayling is a fish of the year's old age: of that time when the morning silver of early winter mingles with the russet and amber of the woods, that want but the midday light of the Enchanter to blossom into gold ...[3]

Between 1890 and 1893 the *Yorkshire Weekly Post* published a series of booklets on Yorkshire rivers. The Crimple did not feature; it is a small stream, apparently lacking the energy to have gouged out a valley that called for a thirty-one-arch viaduct to span it. A more rewarding river for Maskell's fly fishing was the Wharfe, the first in the *Yorkshire Weekly Post*'s series, which was brought within walking distance when the North Eastern Railway transferred him to Arthington junction several years later (Fig. 59). The Wharfe rises where two becks meet in Langstrothdale, thirty-two miles to the north-west. For much of the distance between it flows over limestone. When the water level falls during dry summers, smooth hollows and ridges of white rock are revealed, and the river is said to be 'on its bones'.

59. Maskell Mitchell and his family, c. 1908. Bernard Mitchell stands far right. Mack, who goes to Canada, stands on the right to rear.

In 1906 Maskell left the railway and followed his heart to join a fishing tackle business in Leeds. Leeds is on the Aire, of which the *Weekly Post*'s rivers series said in 1893: 'We hear strange legends of the once pure state of the river and of sea trout being caught at Leeds Bridge.' The 'dark and dirty' Aire now dragged its course through the centre of the town, 'a putrid bed of excrement'.[4] The business was owned by Francis Maximilan Walbran, a dedicated fly fisherman who wrote on fishing for the *Leeds Mercury* under the pen name 'Halcyon', and for the *Field* as 'Wharfedale', and had been a columnist for the *Fishing Gazette* since the 1870s. Like Pritt, he was fascinated by grayling, for whom he devised flies such as the Red Tag and Sea Swallow, and lost his life while fishing for them in the River Ure at West Tanfield in February 1909.[5] It is not clear what happened. The Ure can rise several feet in an hour, and one theory said that he was caught by rising water in dangerous wading ground. Another suggested that his hat had blown off and that in trying to recover it his waders filled and pulled him down as he stumbled off a ledge into deep water.[6] The churchyard at West Tanfield slopes down to the river. Years before, Walbran had told a friend: 'I should like to be laid there ... just within sound of the bubbling river.'[7]

His wish was granted, and Maskell Mitchell took over his business.

Mitchell's move to Leeds meant that he lost the tied house that had gone with the signalman's job, so he moved his now twelve-strong family to a stone-built terraced house in the nearby village of Pool-in-Wharfedale. Road traffic was then light, consisting almost entirely of horse-drawn carts and bicycles. Bernard and his friends accordingly used the road in front of their house as a cricket pitch. The West Riding County Council had not yet got around to improving the surface of rural roads, and when occasional cars did begin to pass the boys learned to jump into neighbouring fields to escape the clouds of white dust which billowed behind them and coated roadside hedges. Dusty hedges were less of a problem than road tar: when the WRCC did begin to spray Pool road, tar washed into the Wharfe and killed the fish.

By 1914 sixteen-year-old Bernard was working as a mechanic in Otley. In May that year he overstated his age to join the Territorial Army. He did so, he said later, because the annual training camp beside Pembrey Sands in Carmarthenshire was a way of getting to stay by the sea for a fortnight.[8] However, his battalion's arrival at Pembrey coincided almost to the hour with the outbreak of the First World War and an immediate recall to Yorkshire. His unit's guns were drawn by horses, and he spent the next few months on Doncaster race course, striving to break in un-cooperative animals that the army had randomly commandeered from local farmers. In June 1917 Bernard was one of seven Otley lads resting in a dugout when a German shell came through its corrugated iron roof. Five of them were blown into such small pieces that their remains were collected by hand and put in sandbags; a sixth died soon afterwards. Bernard spent nine months in hospital, several more in convalescence, was retrained as a gas instructor, and eventually discharged minus his right eye. Bits of shrapnel continued to work their way out of his flesh for years. This was not the kind of holiday he had had in mind three years before.

The *Wharfedale Observer* discovered that Maskell Mitchell had five sons serving in the army. The paper published an illustrated article that described them as the 'Five Pool Brothers'. A copy came into the hands of Robert Bright Marston, editor of the *Fishing Gazette* and a spirited campaigner against careless use of road tar. Marston reprinted the portraits and asked Maskell Mitchell if his sons were anglers. Maskell replied: 'Yes, they are all devotees of the fly rod, and William . . . was on the lower Costa two seasons ago and caught a lovely 3½lb trout on the May Fly. All have been my companions on many fishing expeditions,

and now they are gone the sport has lost a great deal of its charm.'⁹
Costa is the Costa Beck, a small river in Ryedale then noted for the
abundance of its trout and grayling, and the clarity of its waters across
which kingfishers sped like electric-blue sparks. The name has a Scandinavian root; it means something like 'full of life'.¹⁰

The Five Pool Brothers all came home. But home no longer felt the
way it had in the 'great gold Maytime'. Four of them left for Canada.
Last to go was Maskell the younger (known as Mac), who became a
prairie farmer in Saskatchewan. Bernard, meanwhile, had joined his
father in the fishing tackle business in Leeds. Around the same time,
Sir Eric Geddes was appointed Britain's first Minister of Transport.¹¹
One of Geddes's innovations was the introduction of a scheme to cat-
egorise roads according to their importance and allocate funds for their
building and repair. Funding was determined by a road's place in the
Ministry's hierarchy. Roads in the top two classes were to be given a
letter, A or B, which indicated its grade, and a number, which in England and Wales indicated its place in a geographical scheme denoted by
six clockwise-numbered trunk roads that radiated from London.¹² Thus
the A1 ran north from London and roads beginning with 1 are found to
its east, in Essex, East Anglia, Lincolnshire, eastern parts of Yorkshire
and Durham and a sliver of Northumberland. A road that crossed more
than one zone bore the root number of the zone from which it began,
working counter-clockwise. Yorkshire west of the A1 was in Zone 6, so
the road connecting Leeds with Scarborough, say, is the A64.

The interwar upgrading of trunk roads, expansion of the small-car
market and low-priced motor fuel offered West Riding families an al-
ternative to the railway for day trips (Fig. 60).¹³ During the 1930s the
mass production of affordable touring caravans,¹⁴ better cars, improved
couplings to join them and the Holidays With Pay Act brought the
coast, Moors, Wolds and Dales ever more within the range of individual family travel.¹⁵ Distinctive wayside landscapes of pubs, road-houses,
cafés, and filling stations grew up in result. Families bound for resorts
like Filey or Bridlington had favourite stopping places and waypoints to
which they eagerly looked forward.

Just past York there was a café and garage called the Hazelbush,
where Bernard Mitchell's brother-in-law liked to stop with his family
on their way to Scarborough. Next came The Four Alls, a pub so-called
because 'the queen rules all, the bishop prays for all, the soldier fights
for all and John Bull pays for all'. The Four Alls is still there, and with its
tile-hung gables, tall brick chimney and sandy pine-clad setting it might

60. *Heading east: a new trunk road near Hessle, 1931*

be a piece of Surrey accidentally stranded in Yorkshire. Further on was the Tanglewood, formerly the Bide-a-While café, and then the Lobster Pot inn. Next was The Spitalbeck, a road-house that foreshadowed the climb through rolling hills towards Malton. From 1936 your progress now picked up because the next few miles were on new-built dual carriageway. Hearts lifted hereabouts because workaday life was now well and truly left behind. On the way back, the children tired, sand between their toes, thoughts turning to the coming week, The Spitalbeck marked the approach of Sunday's end.

Roads like the A63 from Leeds through Selby towards Hull, or the A166 from York to Bridlington, each had their own individual feel which became extensions of the places to which they led. Since the 1970s those harmonics have largely faded, lost behind road widening, retail parks and estates of phoney Georgian and Victorian houses with names like Barley Fields and Holmes Meadow. Here and there, however, you can still catch them: on stretches that retain the 1920s eighteen-foot width, or where runs of semis and council houses mark the edges of places that holidaymakers knew eighty years ago. The A166 across the Wolds keeps a unique air, stirred through such things as a solitary brick garage near Full Sutton, the long ascent of Bishop Wilton Wold (nemesis of many a charabanc and radiator), reverberant names like Wetwang and Garton Slack, and the vast gently rolling distances and dotted woodlands of

61. Duggleby Howe on the skyline of the Yorkshire Wolds

the plateau (Fig. 61). If you take the lesser road through Sledmere, past
enclosed valleys and parklands, the memorial tower to Sir Tatton Sykes,
erected to his memory in 1865 by those who 'loved him as a friend
and honoured him as a landlord', stands in solitude like a rocket on a
Kazakhstan cosmodrome.

The railways worked hard to compete. Yorkshire was served by two
of the 'Big Four' companies created by the amalgamations of 1921. The
London, Midland and Scottish Railway (LMS) linked western Yorkshire towns with north Wales, the Lake District and Lancashire resorts.
The London and North Eastern Railway (LNER) covered the greater
part of Yorkshire and approaches to its coast. Resorts on both sides of
the Pennines had by then evolved to cater for different classes and pref-
erences, and there were local parks and favourite open spaces weekends
(Fig. 62). Morecambe was frequented by skilled technical and supervis-
ory workers from West Riding towns. Southport, with few pubs and
a lot of Nonconformist chapels, was considered respectable. Blackpool

62. *Shipley Glen, 1938*

was open and jaunty. Alongside them, looking east, Bridlington and above all Scarborough appealed to Lancashire cotton workers. During the Victorian period the weaving towns did much for the development of working-class holidays that lasted for several consecutive days. In places where employment was centralised, if a labour force took a customary holiday by voting with its feet it could do so without reprisal. Such solidarity assisted collective saving through 'going off' clubs, and the application of pressure for the incremental extension of days granted. By the 1890s the result was a system whereby entire towns boarded trains together to go on holiday in staggered rotation. Lancashire cotton workers 'had longer consecutive recognized summer holidays at an earlier date than anywhere else in industrial England; and their observance of a regular working week for the rest of the time made it easier for them to save and prepare for a seaside holiday'.[16] This meant that the demand for seaside visits was spread across July and August, 'as different towns took their holidays at different times; and this accidental stagger effect made it possible for a specialised working-class holiday industry to emerge at an early stage'.[17] It also came to suit the employers, who used the holiday period to service machinery and inspect their chimneys. The Wakes system was imitated in West Riding woollen towns. A similar pattern emerged in Sheffield, where 'well-paid aristocracies' in the steel, cutlery and light metalworking industries went to holiday in Scarborough and Cleethorpes, and Bridlington by the 1900s had come to be known as 'Sheffield-by-the-Sea'.[18]

Sheffield's elite craftsmen remind us that while the working-class holiday became mass experience, the working class was not an undifferentiated mass. The LNER's *Yorkshire Coast Holiday Handbook* and interwar posters emphasised the individuality of Yorkshire's coastal towns. Hornsea with its adjoining mere became 'Lakeland by the Sea'. Filey was 'for the family'. The bold, flat colours of Tom Purvis's posters for Bridlington and 'East Coast Joys' spoke of vigour. As late as the 1960s posters published by the successor region of British Railways depicted Redcar as somewhere rather superior, an idea partly corroborated by houses on its outskirts which point to a flirtation with modernism in the 1930s.

In 1925 Filey's official guide defined the town by what it was not. Prospective visitors were told that they would find 'no stalls, no palmists, no itinerant musicians' in Filey. 'Cheap trippers never come here, and they are not wanted. If they did come they would feel themselves lost. Scarborough to the left and Bridlington to the right cater for the tripper. Filey views him with profound disdain.'[19]

My grandmother was among those whom Filey disdained, which might be why nobody in the family ever mentioned it. On August Bank Holiday 1910, you will recall, she, my great-grandmother and infant mother-to-be were cheap trippers at Scarborough, where all those long curving station platforms had been built to receive trainful after trainful of excursionists.[20] One place that never featured on an LNER poster was Carlin How, to which Hannah Smith and Emily Wearne returned on that Bank Holiday Monday evening. Nor was Skinningrove on Maskell Mitchell's list of fishing places, many of which back in his signalling days had been selected for their proximity to stations of the North Eastern Railway and his consequent ability to visit them on days off. No grayling had gone anywhere near the Skinningrove beck for years.

Seven years after the car on the beach photograph, Mitchell wrote in consolation to Robert Bright Marston, following the death of Marston's son in action. Mitchell assured Marston that his son had died 'a glorious death, fighting for home and freedom'. He then reflected:

> I hope that our other sons will live to come home again, and again
> wander with us by the side of our pretty trout streams. Out of our

little village of Pool over 100 young men have joined the Army. The place seems dead now the boys are gone.

Marston printed Maskell Mitchell's letter in the *Fishing Gazette* on 27 January 1917. On the 30th one of Alf Myers's former neighbours in Carlin How was among the day's 335 fatal Allied casualties. He was Richard Rushmer Cryer, a member of the Yorkshire Regiment who appropriately enough for an ironstone miner had been attached to the 176th Tunnelling Company of the Royal Engineers. Tunnelling was a substantial enterprise on the Western Front. Around 25,000 men en-gaged in it; there were something like seven and half miles of Allied tunnels under Vimy Ridge alone, variously excavated to plant mines under enemy positions, for communication, services, and to counter enemy underground activity. Mine warfare was a realm for terrifying tunnel visions, of entombment, drowning, obliteration and carbon monoxide poisoning. Richard was one of four members of the 176th Company who died in a single incident. He left a wife and six children. His brother, Joseph, was killed eight months later. We do not know if or when Alf Myers was told of his neighbours' deaths, but it is likely that the news did reach him, for the youngest of the Cryer brothers was Charles, and Charles had been another member of the Richmond Sixteen.

The Cryers lived around the corner in Dixon Street. Like Myers, their father, William, was a deputy in the Carlin How mine, where Charles had worked as an onsetter until March 1916, when the appeal tribunal at Guisborough declined to accept that his socialist belief in the unity of all peoples was an obstacle to his participation in the war.[21] For a time Cryer's itinerary went in step with that of Myers – Richmond Castle, Boulogne, the death sentence, commutation to penal servitude, HM Prison Winchester, tribunal at Wormwood Scrubs, Dyce Camp (Fig. 63), the Wakefield Work Centre. After Wakefield, however, on 4 March 1918, Myers went to Maidstone gaol and Cryer was sent to a work centre at Dartmoor prison.

One of many criticisms of the work centres was the pointlessness of their work: instead of tasks that were socially useful, men were frequently assigned to duties that were worthless, disgusting, or for which they were purposely left ill-equipped, or some combination of the three.[22] Yet early in the history of the Dartmoor centre:

the Harmsworth Press thought that there was material there for a popular stunt against the COs. Newspaper men were sent down to

63. Companions together: conscientious objectors at Dyce Camp, autumn 1916. Charles Cryer is third from left in back row.

Princetown for the purpose of taking photographs and seeking interviews. They were admitted by the authorities. Articles were headed, with more regard to alliteration than to accuracy, 'Princetown's Pampered Pets', 'Coddled Conscience Men', 'The COs' Cosy Club' ... Interviews, mainly mythical, were printed.[23]

The ascription of sloth and pleasure-seeking to conscientious objectors was a recurrent theme in the *Daily Mail*, which said that 'Conscientious Shirkers' were the 'exclusive undivided shame of England', motivated by cowardice and – a little puzzlingly in the light of Dyce Camp, or the work centre where men living in one room with an open lavatory and no washing facilities rose at 4.30 a.m. each day and donned the same stinking clothes to break carcasses with a pick and shovel for fertiliser – as 'afraid of discomfort'.[24] The *Mail* described COs as 'unworthy creatures' who were unwilling to defend their womenkind against 'such monsters of cruelty and lust' as Germans.[25] (This view of German character was reflected in the internment of members of Britain's German civilian community, many of whom were held in a camp at Lofthouse Park near Wakefield, another kind of 'Little Germany' where huts a bit like static caravans were available to detainees for a fee. See Fig. 64.[26]) The day after Charles Cryer's brother, Joseph, fell, an editorial headed 'The Revolting "Conchy"' examined the merits of a call from the Bishop of

64. Internment camp, Lofthouse Park, First World War

Exeter for the release of those whose scruples were based on religious principle, and the dispersal of the rest, 'whose attitude towards the war is political (that is to say, malignant) . . . in small groups, without money or rations, throughout the country, and preferably in areas where they are likely to be bombed'.[27]

In the eyes of the *Mail*, then, Charles Cryer and Alf Myers had been on one long, state-subsidised holiday. When I was taken to Carlin How in the early 1950s both of them still lived nearby. Like the rest of us, their ancestors came from all directions, as often as not becoming Yorkshire as being born into it, and their children bearing memories of Yorkshire to faraway destinies with which they became redefined. This continues: about a fifth of Bradford's people have ancestors in Pakistan. The deepest faith of Cryer and Myers was neither in Yorkshire nor in nations, but in the interdependence of the human family. As faiths go it sits well in a place that began near the far end of the world.

The place itself is middling. Grayling are back in the Skinningrove beck; the fast-rising Ure and Wharfe run clear; sea trout have been seen again in the Aire at Leeds. On the other hand, such is the enthusiasm for driven grouse-shooting that you will be lucky to see even a buzzard up Washburndale; the Costa Beck has been designated as a 'failing water body'; and the eastern population of North Atlantic right whales is thought to be functionally extinct. Meanwhile, just short of half of Dewsbury's citizens profess Islam, and descendants of the Wearnes, Mitchells, Myers and Smiths who laboured in mines and mills a century and a half back are spread across North America. The Carlin How they left is still a bit less than a village, a touch more than a hamlet. And if you enter Yorkshire and your world begins, with Daniel Defoe you may still be uncertain which way to begin to take a view of it.

GLOSSARY AND ABBREVIATIONS

Adit	Horizontal tunnel giving entrance to a mine
Bailey	Area enclosed by ditch and palisade adjoining a **motte**
Baleen	Plate with bristly fringes made of keratin, providing basis for filter-feeding in the mouths of baleen whales
Barque	Three-masted sailing vessel, rigged square at the fore and main masts, with a fore-and-aft sail on the mizzen
Bernicia	Region including south-east Scotland and north-east England north of the Tees that was home to a dynastic group in the sixth century and earlier seventh, later merged with **Deira** to form the kingdom of Northumbria
Bill	An infantry polearm weapon
Bluff	Steep slope or cliff created by incised river meander
Brig	Sailing vessel with two square-rigged masts
Carding	Process of separating and straightening wool fibres
CO	Conscientious objector
Cobble, Coble	High-bowed boat used by fishermen in north-east England and eastern Scotland
Combing	Preparation of carded fibre for spinning
Copepod	Small crustacean
Dale	Valley
Deira	Region between the Humber and Tees that was home to and ruled by a dynastic group in the sixth century and earlier seventh, later merged with **Bernicia** to form the kingdom of **Northumbria**
Drift	Horizontal passage in a mine following a bed of mineral or ore

East March	Area of late medieval/early modern Anglo-Scottish border
Flinching	Removal of blubber from a whale carcass
Fother (1)	Staunch a leak in a ship with a sail
Fother (2)	Unit of mass for lead
Francia	Early medieval kingdom initially centred on an area extending eastwards from modern northern France/Belgium to Thuringia and from the Netherlands south to the Franco-German border near the source of the Rhone. At full extent in the ninth century the kingdom and its dependencies covered most of today's France, Benelux, Germany and northern Italy
Ginnel	Narrow passage between buildings at right angles to a town street giving access to back land (cf. chare, wynd, snicket, gulley, twitchel)
Greywacke	Sandstone formed of sand and clay
Gypsy	Stream with sporadic flow
Hurrier	Child or woman harnessed to a small wagon of coal in a mine
Hushing	Method of mining using a torrent of water to expose mineral veins
Ice front	Edge of sea pack ice at a given time
Keel	Cargo-carrying sail craft used in estuaries and inland waterways
LNER	London and North Eastern Railway
Moor	Tract of uncultivated, ill-drained, peaty open land; sometimes heath
Moraine	Accumulation of soil and rock deposited by glacial action
Motte	Tall conical mound surmounted by palisade and tower, built and used between the eleventh and thirteenth centuries
Mungo	Fibrous material recovered from close-woven cloth waste
Northumbria	Early medieval kingdom which at maximum extent in the early ninth century included most of northern England, the Scottish Borders and Lothians
NCC	Non-combatant Corps

Onsetter	One who loads and receives tubs that travel in and out of a mine
RCAF	Royal Canadian Air Force
Schooner	Sailing vessel evolved in the sixteenth/seventeenth century by Dutch shipbuilders with fore-and-aft sails rigged from two or more masts, the main mast being taller than the fore
Shoddy	Wool respun from the fibres of old clothes that have been ground down
Specksioneer	Chief harpooner
Spyhop	Whale behaviour in which the whale thrusts its head vertically above the water surface, steadying itself by use of pectoral fins to obtain a view or better acoustic picture of its surroundings
Stuff	Generic term for woven fabric
Thruster	Child or woman who pushes a small wagon of coal being drawn by a hurrier
Top-making	Process of washing, combing and sorting raw wool which yields longer fibres (tops) that are ready for spinning
Winterbourne	Seasonal stream on chalk or limestone that flows in wet weather
WRCC	West Riding County Council

ACKNOWLEDGEMENTS

For advice, information, and many different kinds of help while this book was being written my thanks go to Jon Aars, Fiona Barnard, Sarah Bastow, Susan Bolton, Kevin Booth, Eileen Burgess, Keith Butcher, Gillian Cookson, Neil Cossons, Bob Cywinski, Glenn Foard, Shannon Fraser, Alan Garner, David Jennings, Keith Laybourn, Rob Light, Tom Lord, William Marshall, Keith Miller, David Neave, Robert Owen, Cyril Pearce, Dominic Powlesland, David Pybus, Mick Sharp, Adam Smith, Mike Spence, David Stocker, Tom Tew, Peter Tuffs, Dave Webster, Roger W. H. Whiteley, Gillian Wood, Ian Wood, and Michael Wood.

For help at archives I thank staff in the Bradford Museums and Galleries, Cleveland Ironstone Mining Museum, Collections and Research Centre at Mystic Connecticut, East Cleveland Image Archive, East Riding Archives and its Flickr channel, English Heritage, Historic England, the universities of Huddersfield, Hull, Leeds and Manchester, Hull History Centre, Hull Maritime Museum, Humber Keel and Sloop Preservation Society, Imperial War Museum, the National Archives, North Yorkshire County Record Office, North York Moors National Park, Norwegian Polar Institute, Whitby Literary and Philosophical Society Whitby Museum, York Archaeological Trust, York Museums Trust, York Press.

For permission to quote from the work of others I thank Faber and Faber for lines from 'Bridge for the Living' by Philip Larkin and 'Moors' from *Remains of Elmet* by Ted Hughes; Virago, for extracts from Winifred Holtby's description of the attack on Scarborough in 1914; Penguin Random House for passages from J. B. Priestley's *Bright Day*; and Elizabeth Winthrop Alsop for the extract from Stewart Alsop's letters.

Warm thanks are given to Amy Levene for her characterful maps in the text, and to Tracey Partida for the map of Yorkshires physical geography (Plate 1).

And last: how do I thank Jane for her mentoring on all things York-shire? When she agreed to marry me nearly half a century ago, if was on condition that I moved to join her in God's Own Country. I hope this book will show her that I understand why.

NOTES

INTRODUCTION:
AVALONIA TO BEMPTON CLIFF

1. This vehicle was registered in the West Riding in 1904.
2. Anthea Jervis, *Dating and Interpreting Victorian Portrait Photographs*, Museums Association, 2009, 6.
3. Weather at the start of that August was cool and changeable: *Monthly Weather Report of the Meteorological Office, August 1910*.
4. 'The loudest shouters often don't have much to sell.'
5. Viðar Hreinsson, ed., *The Complete Sagas of Icelanders*, vol. 1, Reykjavík: Leifur Eiríksson Publishing, 1997.
6. See Chapter 9.
7. Arthur Mee, *Yorkshire North Riding*, London: Hodder & Stoughton, new edition 1970, 185.
8. Winifred Holtby Collection, Hull History Centre.
9. Simon Winchester, *The Map That Changed the World: A Tale of Rocks, Ruin and Redemption*, London: Penguin Books, 2002.
10. *Scarborough Herald*, 5 September 1839, 2.
11. See Chapter 2.
12. Richard Bell, *Yorkshire Rock: A Journey through Time*, Keyworth: British Geological Survey, 1996, 28–9.
13. D. B. Smith, 'The Palaeogeography of the British Zechstein', in L. F. Dellwig and J. L. Rau, eds., *Third Symposium on Salt*, vol. 1, Cleveland, Ohio: Northern Ohio Geological Society, 1970, 20–23.
14. D. B. Smith, 'The Late Permian Palaeogeography of North-East England', *Proceedings of the Yorkshire Geological Society* 47, 1989, 285–312
15. See Chapter 2.
16. G. K. Lott and A. H. Cooper, *The Building Limestones of the Upper Permian, Cadeby Formation (Magnesian Limestone) of Yorkshire*, British Geological Survey Internal Report IR/05/048, Nottingham: NERC, 2005. Other formations in the Zechstein Group are a calcareous mudstone, the Roxby Formation (previously Upper Marl); the Brotherton Formation of dolomitic limestone (formerly Upper Magnesian Limestone); and the Edlington Formation, a calcareous mudstone with gypsum (formerly Middle Marl).
17. Calcaria in the (?) third-century Antonine Itinerary, *Kælcacæstir*, in Bede's *Ecclesiastical History* (*c.* 731), probably coincident with modern Tadcaster.
18. Sarah Browne, 'The Building of the Chapter House', *'Our Magnificent Fabrick': York Minster: An Architectural History* c 1220–1500,

Swindon: English Heritage, 2003, 46–85. The same text was used earlier in the thirteenth century for the new chapter house at Westminster Abbey.

19. The desert sandstone and marl are mostly buried beneath later deposits, but here and there can be glimpsed in the sides of railway cuttings.

20. See Chapters 1, 3, 4 and 11.

21. K. M. Clayton and Allan Straw, *The Geomorphology of the British Isles: Central and Eastern England*, London: Methuen, 1979, 13–14; J. K. Wright, 'The Stratigraphy of the Yorkshire Corallian', *Proceedings of the Yorkshire Geological Society* 39, 1972, 225–66.

22. See Chapters 1 and 2.

23. A section of Yorkshire's coastal strata was painted round the interior of Smith's museum of geology in Scarborough.

24. Sam M. Slater and Charles H. Wellman, 'Middle Jurassic Vegetation Dynamics Based on Quantitative Analysis of Spore/Pollen Assemblages from the Ravenscar Group, North Yorkshire, UK', *Palaeontology*, 59:2, 2016, 305–28; Helen S. Morgans, 'Lower and Middle Jurassic Woods of the Cleveland Basin (North Yorkshire), England', *Palaeontology* 42:2, 1999, 303–26.

25. Albertus Magnus, *Book of Minerals*, trans. Dorothy Wyckoff, Oxford: Clarendon Press, 1967.

26. Michael Leddra, *Time Matters: Geology's Legacy to Scientific Thought*, Washington: Wiley-Blackwell, 2010, 177.

27. Robert Plot, *The Natural History of Oxfordshire*, 1705, 124.

28. Drayton was thinking of seventeenth-century round shot: 'Eight and Twentieth Song', *Poly-Olbion*, 1622, 146. Another common fossil on Yorkshire's Jurassic coast is the belemnite, the skeleton of a squid-like animal which resembles a modern conical bullet.

29. Edward Lovett, 'The Whitby Snake-Ammonite Myth', *Folklore* 16:3, 1905, 333–4.

30. Walter Scott, *Marmion; A Tale of Flodden Field*, Edinburgh: Archibald Constable, 1806.

31. Similar stories were told of Lambton in County Durham and Linton in Roxburghshire.

32. Roger B. J. Benson and Adam S. Smith, *Osteology of Rhomaleosaurus thorntoni (Sauropterygia: Rhomaleosauridae) from the Lower Jurassic (Toarcian) of Northamptonshire, England*, London: Monograph of the Paleontographical Sociey, 168, 2014.

33. R. N. Mortimore, C. J. Wood and R. W. Gallois, *British Upper Cretaceous Stratigraphy*, Geological Conservation Review Series, No. 23, Peterborough: Joint Nature Conservation Society, 2001, 3.

34. Ibid.

35. D. J. A. Evans, C. D. Clark and W. A. Mitchell, 'The Last British Ice Sheet: A Review of the Evidence Utilised in the Compilation of the Glacial Map of Britain, *Earth Science Reviews* 70:3–4, 2005, 253–312.

36. Twenty-four miles of this line work on as the North Yorkshire Moors Railway.

37. See Chapter 9.

THE AINSTY AND YORK

1. P. Sawyer, ed., *Anglo-Saxon Charters: An Annotated List and Bibliography*, London: Royal Historical Society, 1968, No. 1067, 318.

2. *Edweard cyngc* was King Edward the Confessor, ruler of England

from 1042 to 1066. He was writing to Tostig Godwinson (*c.* 1026–66), Earl of Northumbria, and brother of Harold, who ruled England for nine months after Edward's death. Harold died at Hastings resisting the Norman invasion. Tostig beat him to the grave by nineteen days: he was killed while fighting his brother at Stamford Bridge, near York, on 25 September 1066. The 'thegns of Yorkshire' are local landholders, roughly speaking, forerunners of England's gentry.

3. The kingdom included the southern part of Anglo-Saxon Northumbria and adjoining parts of Cumbria. Southern Northumbria in turn had ghosted another kingdom, which back in the seventh century had borne a British name: *Deira*.

4. *Anglo-Saxon Chronicle* (E) 1085.

5. The origin of the nickname is uncertain, but may refer to the finality of the day of doom when all are judged, with no appeal.

1: MY WORLD BEGINS

1. Whence the naming of Steavenson Street: Addison Langhorn Steavenson (1835–1913) was a geologist and chief engineer of Bell Brothers, the company that owned the Carlin How mine.

2. Simon Chapman, *The Loftus Mines Skinningrove*, Guisborough: Peter Tuffs, 1998.

3. J. S. Fletcher, *Picturesque History of Yorkshire*, vol. 6, London: Caxton, 1899, 46; Asa Briggs, *Victorian Cities*, 'Middlesbrough: The Growth of New Community', London: Penguin Books, 1990 241–276.

4. Quoted by Vera Brittain, *Testament of Friendship*, London: Virago, 1980, 46–50.

5. *Daily Mail*, Friday, 18 December 1914, 4.

6. In May 1916 the starting age was lowered to eighteen and the scope extended to married men.

7. *North-Eastern Daily Gazette*, Friday, 16 March 1916; *Whitby Gazette*, 24 March 1916.

8. NRCC/CL, Case Papers 1916–18.

9. Michael Drayton, *Poly-Olbion*, Part 2, 1622, 144.

10. Hansard, 'Conscientious Objectors', House of Commons debate, 10 May 1916, vol. 82, cc.646–50.

11. *Tribunal*, 3, 15 June 1916, 5.

12. 'Conscientious Objectors in France. Death Sentences and Commutations', *Manchester Guardian*, 24 June 1916, 8.

13. Hansard, House of Commons debate, 29 June 1916, vol. 83, cc.1013–18.

14. See p. 225.

15. *The Granite Echo: Organ of the Dyce C.O.'s*, vol. 1, no. 1, October 1916, 2.

16. 'Conscientious Objectors at Dyce', *Manchester Guardian*, 21 September 1916, 6.

17. Hansard, House of Commons debate, 9 October 1916, vol. 86, cc.806–9.

18. Ibid., c.816.

19. 'Conscientious Objectors in Penal Servitude', *Manchester Guardian*, 11 October 1916, 6.

20. 'Two Deaths in Maidstone', *Tribunal*, 16 January 1919, 2.

21. Ann Kramer, *Conscientious Objectors of the First World War: A Determined Resistance*, Barnsley: Pen and Sword, 2013, 150.

22. Hansard, House of Lords debate, 3 April 1919, vol. 34, cc.150–67.

23. Hansard, House of Commons debate, April 1919, vol, 114, cc.1369–70.

24. Typescript recollection (n.d.), University of Leeds Special

Collections, Liddle/WW1/CO/011.

25. David Hey, 'Yorkshire's Southern Boundary', *Northern History* 37, 2000, 31–47, at 44; M. W. Beresford, *The New Towns of the Middle Ages*, Cambridge: Lutterworth Press, 1967, 136; D. Hey, *Packmen, Carriers and Packhorse Roads: Trade and Communications in North Derbyshire and South Yorkshire*, Leicester: Leicester University Press, 1980, 105–13.

26. Daniel Defoe, *A Tour Through the Whole Island of Great Britain*, vol. 3, London, 1742, 100–101.

27. By the 1820s some twelve carrier or stagecoach services ran through Bawtry each week: Philip L. Scowcroft , *By Road and Rail: A Brief History of Tickhill's Transportation*, Tickhill and District Local History Society Occasional Paper 8, 2010, 10.

28. The first bypass opened in 1961.

29. W. H. Auden, 'I like it cold', *House and Garden*, December 1947, 110.

30. See Chapters 7 and 12. Helen Jewell, *The North–South Divide: The Origins of Northern Consciousness in England*, Manchester: Manchester University Press, 1994, 8–25; Dave Russell, 'Defining the North', in *Looking North: Northern England and the National Imagination*, Manchester: Manchester University Press, 2004, 15; A. J. Pollard, *Imagining Robin Hood*, Abingdon: Routledge, 2004, 65–7; A. J. Pollard, 'North, South band Richard III', in Pollard, *The Worlds of Richard III*, Stroud: Tempus, 2001, 45–7; A. J. Pollard, ed., *The North of England in the Age of Richard III*, Stroud: Sutton, 1996, 10.

31. *Landscape Character and Capacity Assessment of Doncaster Borough*, 2007, 168–9.

32. G. Measom, *The Official Illustrated Guide to the Great Northern, Manchester, Sheffield, and Lincolnshire, and Midland Railways*, London: Griffin Bohn, 1861, 420.

33. South Yorkshire Historic Environment Characterisation Project Doncaster Character Zone Descriptions; T. R. Slater, 'Doncaster's Town Plan: An Analysis', in P.C. Buckland, J.R. Magilton and C. Hayfield, *The Archaeology of Doncaster 2. The Medieval and Later Town*, Oxford: British Archaeological Reports British Series 202, 1989, 43–61.

34. Tile maps of the NER may still be seen at the stations of Beverley, Hartlepool, Middlesbrough, Morpeth, Saltburn, Scarborough, Tynemouth, York and Whitby. The tiles were made by Craven Dunnill & Co Ltd of Jackfield in Shropshire: Tony Herbert, *The Jackfield Decorative Tile Industry*, Ironbridge: Ironbridge Gorge Museum Trust, 1978.

35. *Monthly Weather Report of the Meteorological Office*, vol. XXXVI (new series), no. IV, April 1919.

36. 'Obituary', *The Times*, 31 May 1881, 5.

37. Fletcher, *Picturesque History of Yorkshire*, 65.

38. G. A. Wade, 'Towns made by peers', *English Illustrated Magazine*, July 1912, 358.

39. TNA RAIL 1076/38.

40. The line was closed to passenger traffic in 1958. In 1974 the stretch from Saltburn to Boulby was rebuilt to serve the Boulby potash mine.

41. For example: diaries, letters and other sources of Norman Gaudie, Special Collections, University of Leeds: Liddle/WW1/CO/038; Fred J. Murfin, *Prisoners for Peace*, Cornwall Area Meeting of the Religious Society of Friends, 2014; John Hubert Brocklesby, 'Escape from Paganism', from unpublished account. See Catalogue of the

Peace Pledge Union Conscientious Objection Archive: Part 1, First World War.

42. *Report of the Commissioner to Inquire into the operation of the Act and into the state of the population in mining districts*, London: HMSO, 1847, 25. See also Chapter 11.

43. Cudworth 1891, 222G

44. George Werth, cited in Susan Duxbury-Neumann, *Little Germany: A History of Bradford's Germans*, Stroud: Amberley, 2015, 14–15.

45. 'Health of towns – its influence on the moral and physical condition of society', *Bradford Observer*, 24 December 1840.

46. See Chapter 6.

47. 'Rev William Scoresby DD', *The Bradford Antiquary*, 3rd series, 3, 1987, 53–4; Bryan Waites, 'William Scoresby, 1789–1857', *Geographers: Bibliographical Studies*, vol. 4, 1980.

48. Rebecca Fraser, *Charlotte Brontë*, London: Methuen, 1988, 20–23, at 23.

49. Jane, Grace and Sarah were names repeatedly given to Myers girls; Tom, William, Alfred, Henry and General were awarded to boys.

50. Ted K. Bradshaw, 'The Post-Place Community: Contributions to the Debate about the Definition of Community', *Community Development* 39, 2009, 5–16, at 6.

51. D. Neave, 'The identity of the East Riding of Yorkshire', in *Issues of Regional Identity*, ed. E. Royle, Manchester: Manchester University Press, 1998, 184–200, at 187.

52. Ibid., 187–8.

53. Ibid., 188–9.

54. D. Hodgson, *The Official History of Yorkshire County Cricket Club*, Ramsbury: Crowood Press, 1989, 16–17.

55. William Marshall, *The Creation of Yorkshireness: Cultural identities in Yorkshire c. 1850–1918*, doctoral thesis, University of Huddersfield 2011, iii.

56. John Hornsby, *Forty Country Rambles Round Leeds*, Leeds: W. Brierley, 1911.

57. Bogg's *By the Banks of the Wharfe Through Cravenland, Ilkley to Kettlewelldale* (1921) is an example of this kind of recycling.

58. S. A. Cravan, 'The relationship between Edmund Bogg, the Leeds Savage Club and the YRC', *Journal of the Yorkshire Ramblers' Club* 12:5, 1996, 51–4.

59. *Guardian*, Obituary, 6 June 1961; David M. Copeland, 'From Bradford Moor to Silver Dale', doctoral thesis, University of Bradford, 2009.

60. Worsted cloth: see Chapter 11.

61. C. Gordon, *By Gaslight in Winter: A Victorian Family History Through the Magic Lantern*, London: Elm Tree, 1980; Joseph Riley Archive, University of Bradford, collections description, GB 0532 RIJ.

62. 'Readerships and literary cultures 1900–1950', *Newsletter*, December 1913, 3.

63. John Walker Ord, *The History and Antiquities of Cleveland. Comprising the Wapentake of East and West Langbargh, North Riding, County York*, London: Simpkin and Marshall, 1846, 245.

64. R. V. Taylor, *Ecclesiæ Leodienses; or, Historical and Architectural Sketches of the Churches of Leeds and Neighbourhood*, London, 1875; *Anecdotæ Eboracenses. Yorkshire Anecdotes; or, Remarkable Incidents in the Lives of Celebrated Yorkshire Men and Women*, London: Whittaker & Co, 1883.

65. See for example 'Old Yorkshire words', *Leeds Mercury*, 1 March 1884.

66. Within Raistrick's large output were valuable regional surveys, such as *The Pennine Dales* (London: Eyre & Spottiswoode, 1968), which

remain as good a place to approach Yorkshire as any.

67. The North York Moors National Park was designated in 1952, the Yorkshire Dales in 1954.

68. Charlotte Brontë to Ellen Nussey, 5 March 1850, in Juliet Barker, *The Brontës: A Life in Letters*, London: Folio Society, 2006, 286–7.

69. Fraser, *Charlotte Brontë*, 25, referencing Arthur Nicholls to George Smith, 2 April 1857, John Murray Archives.

70. Elizabeth Gaskell, *Life of Charlotte Brontë*, London: Smith, Elder & Co, 1857, Ch. 2.

71. Ibid.

72. The Life of Charlotte Bronte', *The Times*, 25 April 1857, 9.

73. Phyllis Bentley, 'Yorkshire and the Novelist', *Kenyon Review*, 30, 1968, 509–22.

74. J. A. Erskine Stuart, *The Brontë Country: Its Topography, Antiquities and History*, London: Longmans, Green & Company, 1888.

75. The widening of audience has been attributed to Ellis H. Chadwick's *In the Footsteps of the Brontës*, London: Pitman & Sons, 1914: Nicola J. Watson, *The Literary Tourist: Readers and Places in Romantic and Victorian Britain*, Basingstoke: Palgrave Macmillan, 2006, 125.

76. 'Charlotte Brontë's Country', in 'Round about Bradford', *Bradford Observer*, 2 July 1874, 7.

77. Halliwell Sutcliffe's *A Man of the Moors* was advertised as 'A story of the Bronte country': *The Times*, 21 December 1897, 12. 'The Country of the Bronte Sisters' featured on railway posters in the 1920s.

78. *Glasgow Herald*, 28 December 1899.

79. Stuart, *Brontë Country*, 122–6, suggested that different models were used for Thornfield's exterior and interior. The Rydings, a house at Birstall, near Bradford, is a likely candidate for the exterior and provided the model for the 1872 illustration.

80. See Chapter 4.

81. See for example Stuart, *Brontë Country*, 122; *Country Life*, 19 May 1900, 627.

82. Another strand in the story may have derived from the Eyre family at North Lees Hall, Hathersage, where the first mistress of the hall was said to have become demented and was confined to her room, while Gaskell (*Life*, Ch. 28) had heard of a 'governess in a family near Leeds, who married and had a child by a gentleman who was afterwards discovered to be already married, but to a mad wife': note by Sally Shuttleworth in *Jane Eyre*, Oxford: Oxford University Press, 2000, 478.

83. In tandem with his keen eye for writers Smith was also the begetter and proprietor of the successful *Cornhill Magazine* and *Dictionary of National Biography*.

84. *The Times*, 23 April 1891, 9. For Yorkshire enclaves elsewhere in the world: J. Watson, '"Cooked in True Yorkshire Fashion": Regional Identity and English Associational Life in New Zealand before the First World War', in T. Bueltmann, D. T. Gleeson and R. M. MacRaild (eds.), *Locating the English Diaspora, 1500–2010*, Liverpool: Liverpool University Press, 2012, 169–84.

85. R. Church, *A Stroll Before Dark*, London: Bloomsbury, 1965, 168.

86. Philip Larkin, *Collected Poems*, ed. A. Thwaite, London: Faber, 2003, 79, 188.

87. T. Shepherd, *The Lost Towns of the Yorkshire Coast and Other Chapters Bearing upon the Geography of the District*, London: A. Brown & Sons, 1912.

88. Jean Hartley, *Philip Larkin's Hull and East Yorkshire*, Hull: University of Hull Press, 1995, 44.

89. Gaskell, *Life of Charlotte Brontë*, Ch. 1.

90. Raistrick, *Pennine Dales*, 177–8.

91. Ibid., 178.

92. David Hill, *Turner and Leeds: Image of Industry*, Leeds: Northern Arts, 2008, 69.

93. *Leeds from Beeston Hill*, 1816, Yale Center for British Art, New Haven, Conn., USA.

94. Hill, *Turner and Leeds*, 2.

95. The picture entered the private collection of a London wine merchant, but a lithograph made in 1823 gave the scene wider currency: Robert Upstone, 'Allnutt, John (1773–1863)', in Evelyn Joll, Martin Butlin and Luke Herrmann (eds.), *The Oxford Companion to J. M. W. Turner*, Oxford: Oxford University Press, 2001, 5.

96. Jane Sellars, ed., *Atkinson Grimshaw: Painter of Moonlight*, Harrogate: Mercer Art Gallery, 2011.

97. Robert Jones, *Reuben Chappell: Pierhead Painter*, Tiverton: First Light, 2006.

98. C. Morris, ed., *The Illustrated Journeys of Celia Fiennes, 1685–c. 1712*, Exeter: Webb & Bower, 1982, 97–102.

99. Neave, 'Identity of the East Riding', 188; cf. D. Neave, 'Beverley, 1700–1835', in K. J. Allison, ed., *Victoria County History: Yorkshire East Riding*, vol. 6, Oxford: Oxford University Press, 1989, 112–13.

100. J. Markham, ed., *Burton of Beverley*, Beverley: East Yorkshire Local History Society, 1994.

101. John Masefield, *The Bluebells and Other Verse*, New York: Macmillan, 1961, 1; Michelle Pybus, *George Weatherill: His Family and Their Art*, Whitby: Christine Pybus, 2013.

102. Michael Drayton, *Poly-Olbion*, Part 2, 1622, 146.

103. *Guisborough Exchange*, 13 May 1975.

104. For the works railway: C. Shepherd, *Skinningrove Iron and Steel Works: Its History, Railways and Locomotives*, Melton Mowbray: Industrial Railway Society, 2012.

105. C. Killip, in *Skinningrove*, film by Michael Almereyda, Cinema Guild: Survival Media, 2012.

106. TNA, English Indices of Deprivation 2000.

107. C. Killip, *In Flagrante*, London: Secker & Warburg, 1988; cf. Killip, *In Flagrante Two*, Göttingen: Steidl, 2016.

108. S. J. Sherlock, *A Royal Anglo-Saxon Cemetery at Street House, Loftus, North-East Yorkshire*, Hartlepool: Tees Archaeology, 2012.

109. The different views are well summarised and referenced by John Blair, *The Church in Anglo-Saxon Society*, Oxford: Oxford University Press, 2003, 52–3.

110. Ibid., 53.

111. See Chapter 2.

112. Nothing is known about Heiu's background, but her closeness to Bishop Aidan and subsequent installation as founding abbess of a religious house at a place named as Calcaria (usually taken as Tadcaster, near York) imply she was working with royal support: Bede, *Ecclesiastical History of the English People*, ed. and trans. B. Colgrave and R. A. B. Mynors, Oxford: Clarendon Press, 1969, iv.23.

113. Bede, *Ecclesiastical History*, iv.19.

114. As examples: Eanflæd was a member of the Kentish royal family; Hild had East Anglian relatives; a temporary member of the community at Coldingham was Æthelthryth (*c.* 636–79), an East Anglian princess who married the powerful Northumbrian king Ecgfrith (*c.* 645–85) and later

returned to her parental lands to
found a monastery of her own at Ely.

115. B. Yorke, *Nunneries and the Anglo-
Saxon Royal Houses*, London:
Continuum, 2003, 147.

116. A. Breeze, 'Did a Woman Write
the Whitby Life of St Gregory?',
Northern History 49, 2012, 345–50, at
350; Peter Hunter Blair, 'Whitby as
a Centre of Learning in the Seventh
Century', in *Learning and Literature
in Anglo-Saxon England*, ed. Michael
Lapidge and Helmut Gneuss,
Cambridge: Cambridge University
Press, 1985, 3–32; B. Colgrave (ed.
and trans.), *The Earliest Life of
Gregory the Great*, Cambridge:
Cambridge University Press, 1985.

117. The first and last letters of the Greek
alphabet and a title of Christ in
the Book of Revelation. Rosemary
Cramp, *Corpus of Anglo-Saxon Stone
Sculpture*, vol. 1: *County Durham
and Northumberland*, Oxford:
Oxford University Press (for British
Academy), 1984 [1977], 97–101.

118. B. Yorke, *Kings and Kingdoms
of Early Anglo-Saxon England*,
London: Routledge, 1990, 74.

119. P. Hunter Blair, 'The Boundary
between Bernicia and Deira',
Archaeologia Aeliana, 4th series,
27 (1949), 46–59, reprinted in M.
Lapidge and P. Hunter Blair (eds.),
Anglo-Saxon Northumbria, London:
Variorum Reprints, 1984; Yorke,
Kings and Kingdoms, 7.

120. I. Wood, 'Monasteries and the
Geography of Power in the Age of
Bede', *Northern History* 45, 2008,
11–25, at 11–13.

121. R. W. Rix, *The Barbarian North in
Medieval Imagination: Ethnicity,
Legend, and Literature*, Abingdon:
Routledge, 2015, 129.

2: TUNNEL VISIONS

1. Edward Baines, *History, Directory
and Gazetteer of the County of York;
with a variety of commercial, statistical
and professional information: also,
copious lists of the Seats of the Nobility
and Gentry, of Yorkshire*, vol. 2: *East
and North Ridings*, Leeds: *Leeds
Mercury* Office, 1823, 70–72.

2. Viscount Esher, *York: A Study in
Conservation – A Report to the
Minister of Housing and Local
Government*, London: HMSO, 1968,
paras 2.1–2.4.

3. The origins of the scheme at York
went back to 1948: see Bill Fawcett,
'Plan for the City of York (1948)',
York Historian 30, 2013, 1–20. For the
problem generally: Colin Buchanan,
*Traffic in Towns: A Study of the Long
Term Problems of Traffic in Urban
Areas*, London: HMSO, 1963.

4. See for example Hansard, House of
Lords debate on bypass roads and
population centres, 2 December 1971,
vol. 326, cc.567–83.

5. Councillor Douglas Craig, reported
in the *Yorkshire Evening Press*, 26
October 1971.

6. Judy Hillman, 'In for a penny . . .',
Guardian, 14 October 1972, 13.

7. Nathaniel Lichfield and Alan
Proudlove, *Conservation and Traffic:
A Case Study of York*, York: Ebor
Press, 1976.

8. J. B. Whitwell, *The Church Street
Sewer and an Adjacent Building*,
London: Council for British
Archaeology for York Archaeological
Trust, 1976.

9. Arthur MacGregor, *Finds from a
Roman Sewer System and an Adjacent
Building in Church Street*, London:
Council for British Archaeology for
York Archaeological Trust, 1976.

10. P. C. Buckland, *The Environmental
Evidence from the Church Street*

Roman Sewer System, London: Council for British Archaeology for York Archaeological Trust, 1976.

11. E. C. Harris, *Principles of Archaeological Stratigraphy*, London and New York: Academic Press, 1989. There can be exceptions to object–strata relationships: for instance, when objects are moved by sub-surface animal action, or change position as a result of volumetric change or shrinkage during drought.

12. See Chapter 9.

13. W. Hylton Longstaffe, *Richmondshire: Its Ancient Lords and Edifices*, London: George Bell, 1852, 115–16.

14. John Hodgson, *A History of Northumberland*, part 2, vol. 3, 1840, 287; J. Collingwood-Bruce, *Wallet-Book of the Roman Wall*, London: Longman, 1863, 109.

15. Edwin Lankester (ed.), *Memorials of John Ray, Consisting of His Life by Mr Derham . . . with his Itineraries etc*, London: Ray Society, 1846, 149; James A. Sharpe, *Witchcraft in Seventeenth-Century Yorkshire: Accusations and Counter Measures*, York: Borthwick Papers 81, 1992.

16. For example, at Geroldseck and Kyffhäuser (Jacob and Wilhelm Grimm, *Deutsche Sagen*, 1816/1818, nos. 21 and 23), Untersberg, Odenberg (Karl Lyncker, *Deutsche Sagen und Sitten in Hessischen Gauen*, Kassel: Verlag von Oswald Bertram, 1854, no. 6, 5–7), Ållaberg (Herman Hofberg, *Swedish Fairy Tales*, trans. W. H. Myers, Chicago: Belford-Clarke, 1890, 109–10).

17. William of Newburgh, *Historia Rerum Anglicarum*, cap. 28.

18. Alan Garner, *By Seven Firs and Goldenstone: An Account of the Legend of Alderley*, Temenos Academy Papers, 2010; cf. S. Graça da Silva and J. Tehrani, 'Comparative Phylogenetic Analyses Uncover the Ancient Roots of Indo-European Folktales', *Royal Society Open Science* 3:1, 2016, 150645.

19. Eliza Gutch, *County Folk-Lore*, vol. 2: *Examples of Printed Folk-Lore Concerning the North Riding of Yorkshire, York, and the Ainsty*, London: Published for the Folk-Lore Society by David Nutt, 1901, 406–7.

20. Barbarossa's skeleton is said to be in the cathedral at Tyre, his heart and intestines at Tarsus, other tissue in the cathedral at Antioch.

21. Although if a real Arthur was anywhere, he is likely to have been a sixth-century war leader in the north: see Chapter 3, note 22.

22. Miri Rubin, *Mother of God: A History of the Virgin Mary*, New Haven, Conn., and London: Yale University Press, 2009.

23. At over 350 ft (107 m) and 365 ft (111 m), respectively.

24. Michael Drayton, 'The eight and twentieth song', *Poly-Olbion*, 1622, 147.

25. S. W. Cuttriss, 'The Caves and Pot-Holes of Yorkshire', *Ramblers' Club Journal* 1:1, 1899, 54–64.

26. R. H. Tiddeman, *The Work and Problems of the Victoria Cave Exploration*, Leeds: n.p., 1875.

27. Summarised by Tom Lord and John Howard, 'Cave Archaeology in the Yorkshire Dales', in A. Waltham and D. Lowe (eds.), *Caves and Karst in the Yorkshire Dales*, vol. 1: *Buxton*, Buxton: British Cave Research Association, 2013, 239–51, with further references.

28. Ibid., 242.

29. J. A. Gilkes and T. C. Lord, 'A Neolithic Antler Macehead from North End Pot, Ingleton, North Yorkshire', *Transactions of the Hunter Archaeological Society* 17, 1993, 57–9.

30. T. C. Lord, 'An Interim Report on the Discovery of post-Glacial Adult and Juvenile Aurochs, Bos primigenius Bojanus, in a Cave Shaft on the Limestone Uplands of Craven in the Yorkshire Dales; Settle', Lower Winskill Archaeology Centre Occasional Work Notes 3, 1994.

31. J. A. Gilkes and T. C. Lord, 'A Late Neolithic Crevice Burial from Selside, Ribblesdale, North Yorkshire', *Yorkshire Archaeological Journal* 57, 1985, 1–5.

32. The same point applies to periods like the Iron Age and early Middle Ages, when few things appear to have been deposited in caves.

33. Lord and Howard, 'Cave Archaeology', 247; M. J. Dearne and T. C. Lord, *The Romano-British Archaeology of Victoria Cave, Settle*, Oxford: British Archaeological Reports British Series 273, 1998.

34. Lord and Howard, 'Cave Archaeology', 248; R. S. O. Tomlin and M. W. C. Hassall, 'Victoria Cave', in *Roman Britain in 1997, Inscriptions, Britannia* 29, 1998, 439.

35. A. Raistrick and B. Jennings, *A History of Lead Mining in the Pennines*, London: Longmans, 1965; R. White, 'Lead', in R. A. Butlin (ed.), *Historical Atlas of North Yorkshire*, Otley: Westbury Publishing, 2003, 178–80.

36. J. Bayley, 'Non-ferrous Metalworking in Roman Yorkshire', in P. Wilson and J. Price (eds.) *Aspects of Roman Industry in Yorkshire and the North*, Oxford: Oxbow Books, 2002; A. Raistrick, 'A Pig of Lead, with Roman inscription, in the Craven Museum', *Yorkshire Archaeological Journal* 30, 1931, 181–2.

37. White, 'Lead'.

38. William Camden, *Britannia*, trans. Philemon Holland, London, 1610.

The redness was usually caused by spontaneous combustion.

39. Bede, *Ecclesiastical History of the English People*, ed. and trans. B. Colgrave and R. A. B. Mynors, Oxford: Clarendon Press, 1969, i.1, 17.

40. C. Cook, 'Jet', in R. A. Butlin (ed.), *Historical Atlas of North Yorkshire*, Otley: Westbury Publishing, 2003, 189–90; H. Muller and K. Muller, *Whitby Jet*, Oxford: Shire Publications, 2009, esp. 5–13.

41. This realisation is thanks to the work of the North York Moors Caving Club.

42. See Introduction.

3: DERE STREET

1. Henry, Archdeacon of Huntingdon, *Historia Anglorum (The History of the English People)*, ed. and trans. Diana Greenway, Oxford: Clarendon Press, 1996, 22–5.

2. ' ... id est Watlingestrate, Fosse, Hikenildestrate, Herningestrate, quorum duo in longitudinem regni, alii duo in latitudinem distenduntur', *Leges Edwardi Confessoris* (sic) 12; Bruce O'Brien, *God's Peace and King's Peace: The Laws of Edward the Confessor*, Philadelphia: University of Pennsylvania Press, 1999.

3. Geoffrey of Monmouth, *The History of the Kings of Britain*, iii.5.

4. Earninga stræt is recorded in a charter of 955 (BCS 909). The Iccenhilde weg occurs in a document of 903 (BCS 601). The Old English Wæclingastræt appears in a law code of c. 880 (BCS 986)). A strata publica de Fosse is in a charter of 956. Modern archaeology shows that some lengths of some Roman roads may have followed older routes: see p. 66.

5. Traditionally the Icknield Way has

been regarded as prehistoric; modern study shows it to be more recent and post-Roman at least in part: Sarah Harrison, 'The Icknield Way: Some Queries', *Archaeological Journal* 1601:1, 2003, 1–22.

6. C. Babington (ed.), *Polychronicon Ranulphi Higden Monachi Cestrensis*, vol. 11, London: Longman, 1865, book 1, cap. 45, 47.

7. *Great Domesday*, fo. 328v.

8. 'Ros, that dwellith at Ingmanthorpe in Yorkshir a 2. Miles a this side Wetherby, cummith of a Yongger Brother in Descentes tyme part of the House of the Lord Ros': Thomas Hearn (ed.), *Itinerary of John Leland*, vol. 4, Part 1, 1711, 7

9. Burghley GB/NNAF/M170533: 1596 rental; 1648–65 rentals.

10. Susan Wrathmell, *A Review of Historic Buildings Likely to be Affected by the Construction of the A1 Upgrade between Dishforth and Barton, North Yorkshire*, 2005, 4.

11. The B6275.

12. See p. xxii.

13. Castles along the route include Tickhill, Conisbrough, Pontefract and Knaresborough.

14. See p. 79.

15. Castleford (948), Fulford (1066), Stamford Bridge (1066), Northallerton (1138), Myton (1319), Ferrybridge (1461), Towton (1461), Selby (1644), Marston Moor (1644). A strong case has been made for seeking Brunanburh (937) in the same area: Michael Wood, 'Searching for Brunanburh: The Yorkshire Context of the "Great War" of 937', *Yorkshire Archaeological Journal* 85, 2013, 138–59.

16. 4.97 miles (8 km), 1.16 square miles (3 km²).

17. John Leland, *The Itinerary in or about the Years 1535–1543*, ed. L. T. Smith, London: Bell, 1909, vol. 2, art. 4, 27.

18. Colin Haselgrove, ed., *Cartimandua's Capital? The Late Iron Age Royal Site at Stanwick, North Yorkshire, Fieldwork and Analysis 1981–2011*, Research Report 175, York: Council for British Archaeology, 2016, esp. 393–4, 482–3.

19. Tacitus: *Annals*, 12.36, 12.40; *Histories*, 3.45; *Agricola*, 17; Patrick Ottaway, 'The Archaeology of the Roman Period in the Yorkshire Region', in T. G. Manby, Stephen Moorhouse and Patrick Ottaway (eds.), *The Archaeology of Yorkshire*, Leeds: Yorkshire Archaeological Society, 2003, Occasional Paper No. 3, 125–49, esp. 125.

20. P. R. Wilson, *Cataractonium: Roman Catterick and Its Hinterland. Excavations and Research, 1958–1997*, Research Report 128, York: Council for British Archaeology, 2002, 128.

21. Kenneth H. Jackson, *The Gododdin: The Oldest Scottish Poem*, Edinburgh: Edinburgh University Press, 1969.

22. Andrew Breeze, 'Arthur's Battles and the Volcanic Winter of 536–37', *Northern History* 53:2, 2016, 61–172; Neil Whalley, 'The Old North (2009–2015)', http://www.old-north.co.uk/index.html, accessed 26 September 2016; T. M. Charles-Edwards, *Wales and the Britons 350–1064*, Oxford: Oxford University Press, 2013.

23. Fifth-century evidence both for Elmet and for trans-Pennine contact with north Wales is provided by a memorial stone to Aliortus of Elmet, from Llanelhairn in Gwynedd: Christopher Loveluck, 'The Archaeology of post-Roman Yorkshire, AD 400–700', in Manby et al. (eds.), *Archaeology of Yorkshire*, 151–70, at 156–7. See also Andrew Breeze, 'The Kingdom and Name of Elmet', *Northern History* 39, 2002, 151–71.

24. Only a small number of these places have kept their '-in-Elmet' names down to the present, but at least ten are known historically.

25. Nennius, *Historia Brittonum*, 64; Nennius, *Annales Cambriae*, s.a. 616. The Annals say that Ceredic was dead by 616.

26. The tradition attributed the baptism to Rhun, son of Urien, king of Rheged: Nennius, *Historia Brittonum*, 63; Nennius, *Annales Cambriae*, s.a. 626. For discussion of context: David Rollason, *Northumbria, 500–1100: Creation and Destruction of a Kingdom*, Cambridge: Cambridge University Press, 2003, 120–21. Rheged was a British polity that seems to have corresponded with modern Cumbria, parts of south-west Scotland, and Lancashire, and conceivably extended to north-west Yorkshire.

27. Penda's Welsh ally was Cadwallon ap Cadfan (d. 634), king of Gwynedd. See Andrew Breeze, 'Seventh-Century Northumbria and a Poem to Cadwallon', *Northern History* 38, 2001, 145–52.

28. Bede, *Ecclesiastical History of the English People*, ed. and trans. B. Colgrave and R. A. B. Mynors, Oxford: Clarendon Press, 1969, iii.4.

29. Andrew Breeze, 'The Battle of the Uinued and the River Went, Yorkshire', *Northern History* 41:2, 2004, 377–383.

30. F. M. Stenton, *Anglo-Saxon England*, Oxford: Oxford University Press, 1943, 339; Michael Wood, 'Searching for Brunanburh: The Yorkshire Context of the "Great War" of 937', *Yorkshire Archaeological Journal* 85 (2013), 138–59. The battle cluster includes Castleford (955), Bramham Moor (1408), Ferrybridge (1461), Towton (1461) and Seacroft Moor (1643).

31. Ottaway, 'The Archaeology of the Roman Period in the Yorkshire Region', in *Archaeology of Yorkshire*, 125–49, at 125.

32. They include henges and cursus at Scorton and Catterick (River Swale), henges at Thornborough, Nunwick, Hutton Moor, Cana and the Devil's Arrows (River Ure), and henges at Newton Kyme (River Whare) and Ferrybridge (River Aire): *Prehistoric Monuments in the A1 Corridor*, York: Council for British Archaeology, 2013; Alison Deegan, *Yorkshire Henges and Their Environs*, Air Photo Mapping Project, Historic England, 2013.

33. Walter Scott, *Ivanhoe*, 1820.

34. Legio IX Hispana was based at York until the early second century. The latest dated reference to it in Britain is an inscription of 107–8 (RIB 1, 665); it was gone when work started on Hadrian's Wall, with no sign of when, where or why it disappeared. Archaeological finds and possible epigraphic evidence attest the presence of detachments on the lower Rhine, and some of the unit's officers are recorded in different posts at later dates. Historians have speculated about the legion's loss in Judaea (Second Jewish Revolt, AD 132–5) or Armenia. Some recent discussion has revived the older idea that the legion was destroyed in the course of a revolt in Britain around AD 120. See Nigel Pollard and Joanne Berry, *The Complete Roman Legions*, London: Thames and Hudson, 2015, 95–8, esp. 98.

35. See Chapter 10.

36. *Kingdome's Weekly Intelligencer*, 30 April 1644, issue 53.

37. Sources of these attacks included Pontefract, Cawood and Scarborough.

38. *Perfect Diurnall*, 3–10 June 1644, issue 45.

39. Ibid.

40. Lauren McIntyre and Graham Bruce, 'Excavation All Saints: A Medieval Church Rediscovered', *Current Archaeology* 245, 2010, 30–37.

41. Following change from double- to single-track working in 1972, the signal box was reduced to control of the adjoining level crossing.

42. *A more exact relation of the late battell neer York fought by the English and Scotch forces against Prince Rupert and the Marquess of Newcastle*, London: M. Simmons for H Overton, 1644, BL Early English Books, 1641–1700, 228: E.2, no. 14.

43. John Barratt, *The Battle for York: Marston Moor 1644*, Stroud: Tempus, 2002, 108–9.

44. Ibid., 109.

45. Glenn Foard and Richard Morris, *The Archaeology of English Battlefields*, Research Report 168, York: Council for British Archaeology, 2012, 154–5.

46. Not least between Rupert and John, Lord Byron, whose right wing had been bettered by Cromwell during the first half-hour of the battle.

47. *Perfect Diurnall*, 8–15 July 1644, issue 50.

48. Barratt estimates 3,500 horse and a few hundred foot: *Battle for York*, 146.

NORTH

1. Allertonshire, Birdforth, Bulmer, Gilling East, Gilling West, Halikeld, Hang East, Hang West, Langbaurgh East, Langbaurgh West, Pickering Lythe, Ryedale, Whitby Strand.

4: SPYALL, COWTONS AND THE WOUNDS OF CHRIST

1. Cuthbert Sharp (ed.), *Memorials of the Rebellion of 1569*, London: John Bowyer Nichols & Son, 1840, 12–13.

2. Ibid.

3. Secretaries of State: State Papers Domestic, Edward VI–James I (Addenda): TNA SP 15/14, nos. 90, 94, 98, 99; Krista J. Kesselring, *The Northern Rebellion of 1569: Faith, Politics, and Protest in Elizabethan England*, Basingstoke: Palgrave Macmillan, 2010, 55.

4. Secretaries of State: State Papers Domestic, Edward VI–James I (Addenda): TNA SP 15/15, no. 4 (1).

5. Sharp, *Memorials*, 13–15.

6. A region of the Scottish Marches: the Anglo-Scottish border in the later medieval and early modern period.

7. Kesselring, *Northern Rebellion*, 46–7.

8. See Chapter 3.

9. Paul Everson and David Stocker, 'The Straight and Narrow Way: Fenland Causeways and the Conversion of the Landscape in the Witham Valley, Lincolnshire', in *The Cross Goes North: Processes of Conversion in Northern Europe, AD 300–1300*, ed. Martin Carver, Woodbridge: Boydell, 2002, 271–88.

10. This is more or less what the church guide says: 'the present structure was built during the 1450s', *Church of St Mary South Cowton*, Churches Conservation Trust. Pevsner thought likewise: 'Perp entirely', in *Yorkshire: The North Riding (The Buildings of England)*, Harmondsworth: Penguin Books, 1966, 349.

11. '*Orate pro Anima Ricardi Conyers et Alicie uxoris suae*' ('Pray for the souls of Richard Conyers and his wife Alice'). For discussion of the arms and inscription see Alison James, 'To Knowe a Gentilman': Men and Gentry Culture in Fifteenth-Century Yorkshire, PhD thesis, University of York, 2012, 165.

12. See Chapter 9 p. 168.

13. Sir Richard's armour is of the period 1460–90, his hair after 1470, the Tudor livery collar after 1485, and the sword belongs to the late fifteenth century. The ladies' clothes, hair and headdress are characteristic of the years 1475–90. Pevsner (*North Riding*, 350) and others have identified the male effigy as that of Sir Christopher Boynton. However, as James points out, the best Boynton candidate died in 1451, which is too early for the stylistic evidence and contradicted by the Tudor livery collar (James, 'To Knowe a Gentilman', 173–4). Arthur Gardner identified the effigy as Richard Conyers but misplaced the date of his death (giving 1493 instead of 1502): A. Gardner, *Alabaster Tombs of the pre-Reformation Period in England*, Cambridge: Cambridge University Press, 1940, Pl. 225.

14. Unbound hair usually denotes a maiden. Alison James suggests the second effigy could represent a close relative, perhaps a daughter: 'To Knowe a Gentilman', 174. The features of the two ladies are alike, which suggests that the craftsman who carved them was producing a type rather than portraits.

15. *Testamenta Eboracensia*, vol. 3, Surtees Society 45, 1865, 291; William Page, *Yorkshire Chantry Surveys*, vol. 1, Surtees Society 91, 1894, 145.

16. James, 'To Knowe a Gentilman', 168.

17. Other examples of tombs and commemorative activity of this kind in Yorkshire include the Redman and Gascoigne effigies at Harewood; the Cresacres at Barnburgh near Doncaster; Waterton and Welles at Methley; Wytham at Sheriff Hutton. See Pauline Routh, *Medieval Effigial Alabaster Tombs in Yorkshire*, Ispwich: Boydell, 1976.

18. For Richmondshire families: T. D. Whitaker, *A History of Richmondshire in the North Riding of Yorkshire*, 2 vols., 1832; A. J. Pollard, 'The Richmondshire Community of Gentry during the Wars of the Roses', in Charles Ross (ed.), *Patronage, Pedigree and Power in Later Medieval England*, Stroud: Sutton, 1979, 37–59.

19. A. J. Pollard, *The Middleham Connection: Richard III and Richmondshire 1471–1485*, Middleham: Old School Arts Workshop, 1983, 5.

20. *Mountfort v. Conyers*, 1486–1493, TNA C 1/102/21.

21. Calendar of Patent Rolls Henry VII, vol. 1, 92.

22. James, 'To Knowe a Gentilman', 172.

23. Ibid., 162–3.

24. Ibid., ch. 4.

25. British Library, Additional Charters 66451.

26. Anthony Emery, *Greater Medieval Houses of England and Wales*, vol. 1: *Northern England*, Cambridge: Cambridge University Press, 1996, 398.

27. *Testamenta Eboracensia*, 291.

28. Robert Surtees, *The History and Antiquities of the County Palatine of Durham*, vol. IV, London: Nichols, Son and Bentley, 1840, 101.

29. R. W. Hoyle, *The Pilgrimage of Grace and the Politics of the 1530s*, Oxford: Oxford University Press, 2001, 293.

30. For assessment of the numbers see ibid., 293.

31. Ibid., vii, 423–4; Krista J. Kesselring, 'Participants in the Northern Rising (act. 1569–1570)', *Oxford Dictionary of National Biography*, Oxford: Oxford University Press, 2004.

32. Hoyle, *Pilgrimage of Grace*, 420.

33. Sharp, *Memorials*, 45.

34. Secretaries of State: State Papers Domestic, Edward VI–James 1 (Addenda), TNA: SP 15/15, no. 29.

35. See Chapter 3.
36. Sharp, *Memorials*, 59.
37. The route advised by Sussex ran across the Vale of Mowbray, through the Coxwold–Gilling gap, into the Vale of Pickering, then south through the Howardian Hills and Forest of Galtres.
38. On this day Sussex told Bowes – possibly in expectation that the letter could be intercepted – that reinforcement was expected within two days.
39. Sharp, *Memorials*, 86.
40. Ibid., 92–3.
41. Ibid., 93.
42. Ibid., 94–5.
43. According to Bowes's report to Cecil (14 December), 226 deserters jumped from the walls, of whom thirty-five were killed or injured: ibid., 100.
44. Lords Sussex, Hunsdon and Sir Ralph Sadler to the Council, 12 December 1569: ibid., 98.
45. Bowes to Cecil, 14 December 1569: ibid., 101.
46. Susan Taylor, 'The Crown and the North of England, 1559–70', PhD thesis, University of Manchester, 1981; Kesselring.
47. Kesselring, 'Participants in the Northern Rising'.
48. Sharp, *Memorials*, 132.
49. Sussex to Cecil, 2 January 1570: ibid., 133–5.
50. Bishop of Durham to Cecil, 4 January 1570: ibid., 95.
51. George Bowes to Ralph Bowes, 23 January 1570, ibid., 163.
52. Sussex to Cecil, 1 January 1570: ibid., 130, 131.
53. Kesselring, 'Participants in the Northern Rising'.
54. Kesselring, *Northern Rebellion*, 129–36, and at 131.
55. Ibid., 136.
56. Nigel Saul, 'The Gentry and the Parish', in *The Parish in Late*

Medieval England, ed. Clive Burgess and Eamon Duffy, Harlaxton Medieval Studies, vol. XIV, Donington: Shaun Tyas, 2006, 243–60.

5: NO PLACE LIKE HOME: LACKAWANNA, NASSAGAWEYA AND CASTLE DISMAL

1. John A. Schultz, '"Leaven for the Lump": Canada and Empire Settlement, 1918–1939', in Stephen Constantine (ed.), *Emigrants and Empire: British Settlement in the Dominions between the Wars*, Manchester: University of Manchester Press, 1990, 150.
2. See Chapter 12.
3. These areas included Yorkshire, Lancashire, Wiltshire, Gloucestershire and Scotland: *First Report of Select Committee on Emigration from the UK, 1826–7*, (88), 211.
4. Stephanie Jones, *A Maritime History of the Port of Whitby, 1700–1900*, PhD thesis, University College London, 358; Francis E. Hyde, *Liverpool and the Mersey: The Development of a Port 1700–1970*, David and Charles, Newton Abbot, 1971, 112.
5. Hyde, *Liverpool and the Mersey*, 112.
6. Jones, *Maritime History of the Port of Whitby*, 359.
7. Journal of William Easton, 22 June 1834, Whitby Literary and Philosophical Society archive.
8. C. E. Snow, 'Emigration from Great Britain', in *International Migrations*, vol. 2: *Interpretations*, ed. Walter F. Wilcox, National Bureau of Economic Research, 1931, 239–60, at 239.
9. *Report from the Select Committee on Emigration from the UK*, 1826, iv, 404.
10. G. Clark, 'Farm Wages and

Living Standards in the Industrial Revolution: England, 1670–1869', *Economic History Review*, 54:3, 2001, 477–505; but see also Margaret Lyle, 'Regional Agricultural Wage Variations in Early Nineteenth-Century England', *Agricultural History Review* 55:1, 2007, 95–106.

11. P. J. Perry, 'Where was the "Great Agricultural Depression"? A Geography of Agricultural Bankruptcy in Late Victorian England and Wales', *Agricultural History Review* 30, 1972, 45.

12. Clark, 'Farm Wages and Living Standards', 498.

13. Michael Stainsby, *More Than an Ordinary Man: Life and Society in the Upper Esk Valley, 1830–1910*, Helmsley: North York Moors National Park Authority, 2006, 1–21.

14. Ibid., 10–11.

15. Ibid., 17–18.

16. Lucille H. Campey, *Ignored But Not Forgotten: Canada's English Immigrants*, Toronto: Dundurn Press, 2014, 40–43, at 42.

17. Lucille H. Campey, *Planters, Paupers and Pioneers: English Settlers in Atlantic Canada*, Toronto: Natural Heritage Books, 2010, 38–9.

18. Campey, *Ignored But Not Forgotten*, Toronto: Dundurn Press, 2014, 24.

19. Laura A. Detre, 'Canada's Campaign for Immigrants and the Images in *Canada West* magazine', *Great Plains Quarterly*, Paper 2451, 2004, 113–29, at 11.

20. 'Free Farms of 160 Acres: Of Fertile Prairie Land to Every Settler over 18 Years of Age', *Daily Mail*, 5 April 1904, 1.

21. Detre, 'Canada's Campaign for Immigrants', 113–29.

22. This was displayed in a refrigerated glass case.

23. Unpublished letter from Elizabeth Palliser of Potosi, Grant County, Wisconsin, to George Downs and Elizabeth (née Furniss) of Darley, Nidderdale, 5 February 1867, from Jane Simpson of Low Laithe, North Yorkshire.

24. Alan Rayburn, *Naming Canada: Stories about Canadian Place-Names*, Toronto: University of Toronto Press, 2001.

25. K. A. Hicks, *Malton: Farms to Flying*, Mississauga, Friends of the Mississauga Library System, 2006, xii.

26. Ibid., xiii.

27. Ibid., xiii.

28. Ennis Carradice to Jane Whiteley, 20 October 1921, Whiteley Collection.

29. Ibid., 5 July 1920.

30. Jack Carradice, memoir, n.d., Whiteley Collection.

31. Jack Carradice, memoir, 3.

32. Yorkshire Dales Landscape Character Assessment, 'Wharfedale and Littondale', 1, http://www.yorkshiredales.org.uk/about-the-dales/landscape/landscapecharacterassessment/lca_wharfedale-littondale.pdf.

33. Whiteley Collection.

34. The firm was Henry Alcock Bramley and Co, which also had a mill at Colne in Lancashire. The partnership behind it was dissolved at the end of 1846.

35. See, for example, advertisements in the *Bradford Observer*, 22 August 1850; *Manchester Times*, 27 September 1851; *Leeds Mercury*, 8 February 1873.

36. 'Partnerships dissolved', *Manchester Guardian*, 2 July 1877, 7.

37. 9 October 1919: farm sale, Whiteley Collection.

38. *Wharfedale and Airedale Observer*, 23 December 1919, 2.

39. *Craven Herald*, 12 December 1919, 2.

40. Gedd Martin, 'Yorkshire Settlers in New Brunswick', in *Yorkshire*

Immigrants to Canada: Papers from the Yorkshire 2000 Conference, ed. Paul A. Bogaard, Sackville, NB: Tantramar Heritage Trust, 2012.

41. The total strength of No. 6 Group peaked in October 1944, when it stood at 24,741 all ranks, of whom 6,992 were aircrew. Not all of these were Canadian; some RAF personnel would have been among them: William S. Carter, 'Anglo-Canadian Wartime Relations, 1939–1945: RAF Bomber Command and No. 6 (Canadian) Group', PhD thesis, McMaster University, 1989, 66–7.

42. Brereton Greenhous, Stephen J. Harris, William C. Johnstone and William G.P. Rawling, *The Crucible of War 1939–1945: The Official History of the Royal Canadian Air Force*, vol. 3, Toronto: University of Toronto Press, 1994, 635.

43. In fact, these were Highland cattle.

44. A Temple of Victory, attributed to James Paine, built by *c.* 1770.

45. Cited in Elizabeth Winthrop Alsop, 'My Parents Meet', http://www.elizabethwinthrop.com/2014/05/17/my-parents-meet/

46. *Toronto Globe and Mail*, 13 June 1945.

47. A Gregory Elsley had been deputy lord-lieutenant of the North Riding in the mid-eighteenth century: Yorkshire Archaeological Society, MS 461, Elsley family letters and papers.

48. 420 (RCAF) Squadron Operations Record Book, March 1944, TNA AIR 27/1826.

49. See p. 102.

50. 'Last Call for the Avro Lancaster: From Tiger Force to Derelict on the Alberta Prairie', RCAF news article, 31 March 2014; 'The Canadian Lancasters', Bomber Command Museum of Canada.

6: FARTHEST NORTH

1. Other names include Arctic whale, polar whale, Russian whale, and steeple-top whale.

2. Sequencing and analysis of the bowhead's genome indicates that the species has acquired age-retarding mechanisms to do with resistance to disease and cell repair: M. Keane et al., 'Insights into the Evolution of Longevity from the Bowhead Whale Genome', *Cell Reports*, 10, 6 January 2015, 112–22.

3. J. T. Haldiman and R. J. Tarpley, 'Anatomy and Physiology', in J. Burns, J. J. Montague and C. J. Cowles (eds.): *The Bowhead Whale*, Lawrence, Kan.: Society for Marine Mammalogy, 1993, Special Publication No. 2, 71–156.

4. B. Würsig, E. M. Dorsey, W. J. Richardson and R. S. Wells, 'Feeding, Aerial and Play Behaviour of the Bowhead Whael, *Balaena mysticetus*, Summering in the Beaufort Sea', *Aquatic Mammals* 15:1, 1989, 27–37.

5. For comparison, the blubber layer of a sperm whale averages 5–7 inches' thickness. Blubber thickness varies from season to season, according to nutrition and activity.

6. John Richardson, Charles R. Greene, Charles I. Malme, Denis H. Thomson, *Marine Mammals and Noise*, San Diego: Academic Press, 1995, 161; Susanna B. Blackwell, W. J. Richardson, C. R. Greene, Jr and B. Streever, 'Bowhead Whale (*Balaena mysticetus*) Migration and Calling Behaviour in the Alaskan Beaufort Sea, Autumn 2001–04: An Acoustic Localization Study', *Arctic* 60:3, 2007, 255–70.

7. Doreen Kohlbach, Martin Graeve, Benjamin Lange, Carmen David, Ilka Peeken and Hauke Flores, 'The

Importance of Ice Algae-Produced Carbon in the Central Arctic Ocean Ecosystem: Food Web Relationships Revealed by Lipid and Stable Isotope Analyses', *Limnology & Oceanography* 61:6, 2016, 2027–44.

8. William Scoresby, *An Account of the Arctic Regions, with a History and Description of the Northern Whale-Fishery*, 2 vols., Edinburgh: Archibald and Hurst, Robinson and Co, 1820.

9. William Scoresby, *The Arctic Whaling Journals of William Scoresby the Younger*, vol. II: *The Voyages of 1814, 1815, and 1816*, ed. C. Ian Jackson, London: Ashgate for the Hakluyt Society, 2008, 16 May 1815, 164–7.

10. Ibid., 4 July 1814, 68.

11. Ibid., 16 May 1815, 165–6.

12. Ibid., 16 May 1815, 166.

13. 2.3888 ´10¹⁶ in scientific notation.

14. Scoresby, *Arctic Whaling Journals*, 17 May 1815, 167–8.

15. Scoresby, *Arctic Regions*, vol. 1, 175–81, 546.

16. An error – Scoresby meant 1815.

17. Scoresby, *Arctic Regions*, vol. 1, 546.

18. C. S. Elton, *Animal Ecology*, London: Sidgwick and Jackson, 1927. The first appearance of the term in the *Oxford English Dictionary* was also in 1927.

19. Scoresby, *Arctic Whaling Journals*, 15 July 1816, 275.

20. Scoresby, *Arctic Regions*, vol. II, 486–8.

21. Elizabeth Gaskell, *Sylvia's Lovers*, ed. Francis O'Gorman, Oxford: Oxford University Press, 2014, 16.

22. 'The Arctic Expeditions', *Literary Gazette, and Journal of the Belles Lettres, Arts, Politics Etc*, 50, 3 January 1818, 242–3.

23. The area seems not to have acquired a general name until 'the Yorkshire Moors' were popularised by railway posters, and from 1952 by the creation and naming of the North York Moors National Park. Down to the nineteenth century different parts of the plateau were known by individual moor names, of which 'Blakmore' (Leland) and Blackay Moor (Drayton) were the most widely known.

24. John Playfair, *Illustrations of the Huttonian Theory of the Earth*, Edinburgh: Cadell & Davies, 1802.

25. William Scoresby, 'Experiments and Observations on the Development of Magnetical Properties in Steel and Iron by Percussion. Part 2', *Philosophical Transactions of the Royal Society* 114, 1824, 197–221.

26. See Introduction and Chapter 8.

27. Clements R. Markham, *The Lands of Silence: A History of Arctic and Antarctic Exploration*, Cambridge: Cambridge University Press, 1921, 279–84.

28. Scoresby, *Arctic Whaling Journals*, 7, 8, 9 August 1814, 82–3.

29. In this period, Greenland right whale numbers were high, and in summer many came close inshore or into fjords. Much of the earlier whaling was accordingly done in rowed whaling boats launched from onshore summer bases.

30. The principal whaling ports at this point were Hull, London, Aberdeen, Leith and Whitby: Scoresby, *Arctic Regions*, vol. II, 120.

31. Louwrens Hacquebord, 'The Hunting of the Greenland Right Whale in Svalbard, Its Interaction with Climate and its Impact on the Marine Ecosystem', *Polar Research* 18:2, 1999, 375–82.

32. Hull's whaling fleet was much the larger of the two Yorkshire fleets. A few voyages were still being made from Hull in the 1850s, when oil prices inflated by the Crimean War encouraged belief that northern

whaling could again be commercially viable.

33. Hacquebord, 'Hunting of the Greenland Right Whale', 379.

34. Ibid.; D. A. Woodby and and D. B. Botkin, 'Stock Sizes Prior to Commercial Whaling', in Burns et al. (eds.), *Bowhead Whale*.

35. By the 1820s whale oil lighting was being superseded by coal gas. Scoresby argued for the extraction of gas from whale oil on the grounds that its combustion would be chemically kinder to household furnishings, fittings and paintings: *Arctic Regions*, 421–3.

36. Hacquebord, 'Hunting of the Greenland Right Whale', 379, 381.

37. Scoresby, *Arctic Regions*, vol. II, 420–21.

38. Scoresby, *Arctic Regions*, vol. I, 472.

39. Peter Adamson, *The Great Whale to Snare: The Whaling Trade of Hull*, Hull: City of Kingston upon Hull Museums and Art Galleries, n.d., 10.

40. Scoresby, *Arctic Regions*, 241–54.

41. See above, p. 121.

42. Scoresby, *Arctic Regions*, 292–304.

43. The jawbones on display today are from a whale taken in Alaskan waters.

7: HUMBER

1. Winifred Holtby to Jean McWilliam, 21 October 1926, *Letters to a Friend*, eds. Alice Holtby and Jean McWilliam, London: Collins, 1937, 429.

2. Hull History Centre, L WH/8/8.22/01a.

3. Philip Larkin, 'Here', *Collected Poems*, London: Faber and Faber, 2003, 79. The poem opened Larkin's 1965 collection, *The Whitsun Weddings*.

4. TNA MPC 1/56.

5. 'Thorne Moors' is a shorthand that embraces Thorne Waste, Crowle Moor, Rawcliffe Moor, Inkle Moor, Snaith and Cowick Moor, and Goole Moor.

6. Peter Hunter Blair, 'The Northumbrians and Their Southern Frontier', *Archaeologia Aeliana* 4th series, 26, 1948, 98–126. The *Anglo-Saxon Chronicle* refers to Southumbrians (697) and Southumbria (702), apparently indicating northern Mercia.

7. D. N. Riley, 'The End of Ermine Street at the South Shore of the Humber', *Britannia* 5, 1974, 375–7.

8. There were ferries further east – for example, at Whitgift.

9. 8 October 1828: *York Herald and General Advertiser*, 11 October 1828.

10. P. J. Smart, B. D. Wheeler and A. J. Willis, 'Plants and Peat Cuttings: historical ecology of a much-exploited peatland – Thorne Waste, Yorkshire, UK', *New Phytologist* 104, 1986, 731–48, at 731.

11. John Leland, *The Itinerary in or about the Years 1535–1543*, ed. L. T. Smith, London: Bell, 1909, vol. 1, 37.

12. David Hey, 'The Transformation of Inclesmoor', in *A History of the South Yorkshire Countryside*, Barnsley: Pen and Sword, 2015, 79–86.

13. Martin Taylor, *Thorne Mere and the Old River Don*, York: Ebor Press, 1987.

14. House of Commons debate, 19 June 1972, vol. 839, c.136. The recommendation for an airport at Thorne dated from a report in 1967 on behalf of the Consultative Council for Airport Development for the region: 'Airports for the north-east', *Flight*, 17 August 1967, 253.

15. Peter Skidmore, Martin Lambert and Brian C. Eversham, *The Insects*

of Thorne Moors, Sheffield: Sorby Natural History Society, 1987.

16. R. Giblett, *Postmodern Wetlands, Culture, History, Ecology*, Edinburgh: Edinburgh University Press, 1996; C. Firth, *Nine Hundred Years of the Don Fishery: Domesday to the Dawn of the New Millennium*, Leeds: Environment Agency, 1997; cf. Ian Rotherham and Keith Harrison, 'History and Ecology in the Reconstruction of the South Yorkshire Fens: Past, Present and Future', in Bella Davies and Stewart Thompson, eds., *Water and the Landscape: The Landscape Ecology of Freshwater Ecosystems*, Oxford: IALE UK, 2006, 8–16; Ian Rotherham, *The Lost Fens: England's Greatest Ecological Disaster*, Stroud: The History Press, 2013, 54–70.

17. Larkin, 'The Whitsun Weddings', *Collected Poems*, 92.

18. Paul Hughes, 'Roger of Howden's Sailing Directions for the English Coast', *Historical Research* 85 no. 230, 2012, 576–96.

19. Wallingfen enclosure, *Journal of the House of Commons*, 21 March 1777, 290.

20. Martin Millett and Sean McGrail, 'The Archaeology of the Hasholme Logboat', *Archaeological Journal* 144, 1987, 69–155.

21. Joan Liversidge, D. J. Smith, I. M. Stead and Valery Rigby, 'Brantingham Roman Villa: Discoveries in 1962', *Britannia* 4, 1973, 84–106.

22. Martin Henig, *Religion in Roman Britain*, London: Batsford, 1984, 163.

23. Larkin, 'Friday Night in the Royal Station Hotel', 'Here', *Collected Poems*, 130, 79.

24. H. V. Morton, *The Call of England*, fourth edition, London: Methuen, 1929, 17.

25. Jean Hartley, *Philip Larkin's Hull and East Yorkshire*, Hull: University of Hull and Hutton Press, 1995, 6.

26. Hartley, *Philip Larkin's Hull*, 6–7.

27. David Hey, *The Grass Roots of English History: Local Societies in England before the Industrial Revolution*, London: Bloomsbury, 2016.

28. *Hull Packet*, 9 December 1858.

29. Florida, Illinois, Quebec, Iowa, Georgia, Alabama, Massachusetts, Minnesota, North Dakota, Ohio, Texas and Wisconsin.

30. Larkin 'Bridge for the Living', *Collected Poems*, 188.

31. Nikolaus Pevsner, *Yorkshire: York and the East Riding* (*The Buildings of England*), Harmondsworth: Penguin Books, 1972, 322.

32. See pp. 177, 179.

33. *Odd* is from Old Norse *oddr*, 'a spit, point of land'.

34. George Sheeran, *Medieval Yorkshire Towns: People, Buildings and Spaces*, Edinburgh: Edinburgh University Press, 1998, 23–4.

35. Historic England, List entry number 1161853.

36. Historic England, List entry number 1083508.

37. Bricks fired at high temperature, giving darker colour and distorted shape.

38. Larkin, 'Bridge for the Living', *Collected Poems*, 188.

39. The earliest recorded lighthouse was mentioned in 1427. Two lighthouses existed in the sixteenth century. John Smeaton designed two lighthouses, originally lit by coal, later by oil, that were built from 1767 to 1776. A new low light was provided in 1852 and still stands; Smeaton's high light was replaced by the existing lighthouse in 1895.

40. G. de Boer, 'Spurn Head: its history and evolution', *Transactions of the Institute of British Geographers*, 34, 1964, 71–89.

41. George de Boer's model of a 250-year cycle of breakdown followed by regrowth in a new position held sway for many years, but was challenged in 1992 by the Institute for Estuarine and Coastal Studies, which argued that Spurn's present form is a consequence of breaches following nineteenth-century construction and extraction. For reassessment, fresh interpretation and references see V. J. May, 'Spurn Head', *Geological Conservation Review* 28: *Coastal Geomorphology of Great Britain*, ch. 8: 'Sand spits and tombolos', 1–8.

42. Alcuin, *Life of Willibrord*, 1.

43. A lifeboat station has existed on Spurn since 1810.

44. Jan Crowther, *The People Along the Sand: The Spurn Peninsula and Kilnsea. A History, 1800–2000*, Andover: Phillimore, 2006.

45. Phil Mathison, *The Spurn Gravel Trade*, Newport East Yorkshire: Dead Good Publications, 2008, 1.

46. Thomas Sheppard, *The Lost Towns of the Yorkshire Coast*, London: Brown and Sons, 1912, 97–111. A few pieces of the church were retrieved and are still to be seen, embedded in the new church and dotted through village gardens and outbuildings.

47. Larkin, 'Bridge for the Living', *Collected Poems*, 188.

48. Michael Drayton, 'The Eight and Twentieth Song', *Poly-Olbion*, 1621, 149.

49. Israel Finestein, 'The Jews in Hull, between 1766 and 1880', *Jewish Historical Studies* 35, 1996–8, 33–91.

8: WONDERFUL AMY

1. Winifred Holtby, *South Riding*, London: Collins, 1936, 575.

2. Holtby to Jean McWilliam, 10 April 1934, *Letters to a Friend*, 459. Holtby had also been working on *Women and a Changing Civilisation* (1934), which includes reference to women and long-distance flying, for which see Liz Millward, *Women in British Imperial Airspace, 1922–1937*, Montreal: McGill–Queen's University Press, 2007.

3. Hedon (Hull) Aerodrome Committee, Hull History Centre, C TCAE.

4. Winifred Holtby to Jean McWilliam, 10 April 1934, in *Letters to a Friend*, London: Collins, 1937, 459.

5. *Hull Daily Mail*, 1 April 1931.

6. Part of the airport's archive concerns its grazing rights: Hedon (Hull) Aerodrome Committee, Hull History Centre, C TCAE/6.

7. Air Ministry aerodrome map No. 1099, October 1934.

8. John Myerscough, 'Airport Provision in the Inter-War Years', *Journal of Contemporary History* 20:1, 1985, 41–70.

9. 'British and U.S. Air Development: Sir S. Brancker's Comparison', *The Times*, 14 April 1930, 17.

10. *Flight*, 25 January 1934, 75.

11. *Air Transport News*, 24 November 1933.

12. For a time domestic connections were provided or intended with Southampton (via Croydon) and Nottingham: *Flight*, 7 June 1934, 563.

13. 'Hull Airport Greets Her First Big Liner', *Hull Daily Mail*, 31 May 1934, 1; *Flight*, 7 June 1934, 563.

14. *The Times*, 26 February 1935, 19; Provincial Airways poster-timetable, 1935.

15. *Flight*, 2 May 1935, 482.

16. *The Times*, 22 August 1932, 7.

17. *Flight*, 7 September 1933, 896.

18. Ibid., 25 July 1935, 113.

19. *Amy Johnson at Hedon* (1930),

Yorkshire Film Archive, film ID 1101; *Amy's Homecoming 1930*, British Pathé 719.12.

20. *Daily Mail*, 11 August 1930, 7.

21. Ibid., 12 August 1930, 9.

22. Lawrence Wright (1888–1964) was already known as composer of 'Are we downhearted? No!' and as founder in 1926 of *Melody Maker*. He often worked under the pen name Horatio Nicholls.

23. Hohenrein Collection, Hull History Centre.

24. *Daily Mail*, 12 August 1930, 10.

25. The *Mail* first offered a prize to encourage aviation in 1907.

26. *Daily Mail*: 23 May 1930, 7, 11, 13; 25 May 1930, 11; 26 May 1930, 1; 'Miss Amy Johnson's Great Tour in Britain. How Public Will See Her and the "Plane"', 23 July 1930, 1.

27. In 1981 Abrahams was celebrated in the film *Chariots of Fire*.

28. Charles Dixon, *Amy Johnson – Lone Girl Flyer*, London: Sampson, Law, Marston & Co., 1930, 119–20.

29. 'Miss Amy Johnson's Air Tour Plans', *Daily Mail*, 25 August 1930, 7.

30. *Daily Mail*, 7 August 1930, 4.

31. Bernhard Rieger, *Technology and the Culture of Modernity in Britain and Germany 1890–1945*, Cambridge: Cambridge University Press, 2005, 116–57, at 119.

32. 'Miss Johnson's Flight: Arrival in Burma', *The Times*, 14 May 1930, 16.

33. Letter from John William Johnson to Amy Johnson, 10 March 1930.

34. *Daily Mail*, 5 September 1930, 7.

35. Ibid., 7 August 1930, 4.

36. 'Andrew Johnson, Knudtzon Ltd., Importers, Exporters, Commission Agents, Fish Merchants and Salt Fish Curers'.

37. *Report of the R.101 Inquiry*, HMSO, March 1931.

38. D. H. Middleton, *Airspeed: The Company and Its Aeroplanes*, Lavenham: Terence Dalton, 1981, 9.

39. *Flight*, 17 November 1932, 1088.

40. Middleton, *Airspeed*, 12. In October the Tern came second in a cross-country contest for the Wakefield Trophy: *The Times*, 5 October 1931, 16.

41. Blackburn's 1910 premises were in old stables on Balm Road, Hunslet, in Leeds. In 1914 The Blackburn Aeroplane and Motor Company moved into the Olympia Works on Roundhay Road.

42. Boultbee's design work for Handley Page included the Heyford heavy bomber and the H.P. 42 airliner which became the pre-war workhorse for Imperial Airways: *Flight*, 14 September 1967, 427.

43. The prototype Coupé had been built in premises in Burton-upon-Trent, the home town of Boultbee's wife: Eduard F. Winkler, *A Civilian Affair: A Brief History of the Civilian Aircraft Company of Hedon*, Ottringham: Flight Recorder Publications, 2003, 3–8.

44. *Daily Mail*, 20 August 1930.

45. J. A. D. Ackroyd, 'Sir George Cayley: the Invention of the Aeroplane near Scarborough at the time of Trafalgar', *Journal of Aeronautical History*, Paper 2011/6, 130–81.

46. Ibid., 157–8.

47. Ibid., 176.

48. Richard Dee, *The Man Who Discovered Flight: George Cayley and the First Airplane*, Toronto: McLelland and Stewart, 2007, 177.

9: A WET DAY IN BAWTRY

1 'The Pricke of Conscience or the Fifteen Signs of Doom Window in the Church of All Saints, North Street, York', *Vidimus* 45 (November 2010), Features page (ISSN 1752-0741, accessed November 2016).

2. The underlying source for the signs and images was the Book of Revelation, mediated through sources like *The Golden Legend*, a compilation of *c.* 1260: C. Davidson, 'The Signs of Doomsday in Drama and Art', *Historical Reflections* 26, 2000, 223–45.

3. The twelfth and thirteenth scenes in the window reverse the corresponding episodes in the poem.

4. 'Prickle of Conscience', lines 4812–15.

5. See p. 83.

6. Peter Hennessy, *The Secret State: Whitehall and the Cold War*, London: Penguin Books, 2003, 4.

7. It was said that a murdered excise man had been buried beneath the inn's hearth in the 1730s, and that the fire was kept going to ensure that no one would find him. After some decades the always-burning fire became a custom, and an idea took hold that disaster would ensue if it went out. The inn closed in 2007; global financial crisis ensued.

8. See Chapter 1.

9. Bede, *Ecclesiastical History of the English People*, ed. and trans. B. Colgrave and R. A. B. Mynors, Oxford: Clarendon Press, 1969, iii.24, 25; ibid., iv.23.

10. Ibid., iii.25.

11. Ibid., iv.23; Peter Hunter Blair, 'Whitby as a centre of learning in the seventh century', in *Learning and Literature in Anglo-Saxon England*, eds Michael Lapidge and Helmut Gneuss, Cambridge: Cambridge University Press, 1985, 3–32, at 9–11.

12. R. Agar, 'Post-Glacial Erosion of the North Yorkshire Coast from the Tees Estuary to Ravenscar', *Proceedings of the Yorkshire Geological Society* 32, 1960, 409–28. Agar's estimate in 1960 of 300 metres has since been extended to about 500 metres.

13. R. G. Collingwood, 'Roman Signal-Stations on the Cumberland Coast', *Transactions of the Cumberland and Westmorland Archaeological and Architectural Society*, new series, 39, 1929, 138–65, at 143.

14. RIB 721. R. G. Goodchild, 'The Ravenscar Inscription', *Antiquaries Journal* 32, 1952, 185–8.

15. It has been suggested that subsequent storeys were stepped inwards (a configuration known from Roman illustration and survival elsewhere) and for reasons of intervisibility must have risen to about 150 feet: T. W. Bell, 'A Roman signal station at Whitby', *Archaeological Journal* 155, 1998, 303–13.

16. Notably Books 27.VIII.1–9 and 28 of the History by the fourth-century soldier and historian Ammianus Marcellinus (d. after 391); J. G. F. Hind, 'Watchtowers and Fortlets of the North Yorkshire Coast (Turres et Castra)', *Yorkshire Archaeological Journal* 77, 2005 17–24.

17. For example: sources speak of a Pictish assault in 381 which was driven off by the Roman army. On this occasion the army was led by Magnus Maximus who had served under Theodosius in 368, was later granted military command in Britain and went on to be western emperor (383–8). Ottaway notes that the coin series from Filey could fit with the terms of any of Valentinian (367–75), Magnus Maximus (383–8) or even Theodosius I (*c.* 390). For earlier discussion of dating: M. R. Hull 'The pottery from the Roman signal-stations on the Yorkshire coast', *Archaeological Journal* 89, 1932, 220–53.

18. This would fit with representations of seagoing vessels on Pictish symbol stones: Iain Fraser, *The Pictish Symbol*

Stones of Scotland, Edinburgh: Royal Commission on the Ancient and Historical Monuments of Scotland, 2008.

19. Huntcliff (1911/12), Goldsborough 1918; Filey 1857, 1923/29, and 1993–4.

20. Patrick Ottaway et al, 'Excavations on the Site of the Roman Signal Station at Carr Naze, Filey, 1993–94', *Archaeological Journal* 157, 2000, 79–199.

21. Ottaway, 'Excavations on the Site of the Roman Signal Station at Carr Naze, Filey', 190.

22. Huntcliff overlooked Saltburn Sands and Cattersty Sands; Goldsborough gives views of Runswick Bay and Sandsend/Lythe (where there was an early medieval beach harbour); a station at Whitby would overlook both Sandsend and the mouth of the Esk; Scarborough and Filey adjoin large beaches.

23. R. J. Clark, 'A Field Study of the Short-Eared Owl, *Asio flammeus* (*Pontoppidan*), in North America', *Wildlife Monographs* 47, 1975, 1–67.

24. Ottaway, 'Excavations on the Site of the Roman Signal Station at Carr Naze, Filey', 175–6.

25. Sancton is the most northerly of a group of early fifth-century burial places centred around the Wash: Catherine Hills and Sam Lucy, *Spong Hill Part IX: Chronology and Synthesis*, McDonald Institute, University of Cambridge, 2013, 314–24.

26. For the contrasting view that such large cemeteries 'may have been the burial grounds of many different communities and could therefore have acted as places where linkages between groups were negotiated and defined' see Howard Williams, 'Cremation in Early Anglo-Saxon England – Past, Present and Future Research', in H-J. Häßler,

ed., *Studien zur Sachsenforchung* 15, Oldenburg: Isensee, 2005, 533–49; and Howard Williams, 'Cemeteries as Central Places – Place and Identity in Migration Period Eastern England', in B. Hårdh and L. Larsson, eds., *Central Places in the Migration and Merovingian Periods*, Lund: Uppåkrastudier 6, 2002, 341–62.

27. Peter Sawyer, 'The Wealth of England in the 11th Century', *Transactions of the Royal Historical Society* 5 series, 15, 1965, 160; *The Wealth of Anglo-Saxon England*, Oxford: Oxford University Press, 2013.

28. See Introduction, p. xxi.

29. Bede, *Ecclesiastical History of the English People*, ed. and trans. B. Colgrave and R. A. B. Mynors, Oxford: Clarendon Press, 1969, i.1; Martin Daniel, 'A Note on the Anglo-Saxon Lead Industry of the Peak', *Bulletin Peak District Mines Historical Society* 7:6, 1980, 339–41; P. Claughton, 'Lead and Silver: Britain, France and Beyond', *Historical Metallurgy* 42, 2010, 2; P. Claughton, 'Production and Economic Impact: Northern Pennine (English) Silver in the 12th Century', *Proceedings of the 6th International Mining History Congress*, Akabira, 2003, 146–9.

30. Gareth Williams and Barry Ager, *The Vale of York Hoard*, London: British Museum Press, 2010.

31. 'Thomas Baskerville's Journeys in England, Temp. Car. II', in *Historical Manuscripts Commission, Thirteenth Report, Appendix, Part II: The Manuscripts of His Grace the Duke of Portland: Preserved at Welbeck Abbey*, London: HMSO, 1893, 313–14.

32. Ibid.

33. Daniel Defoe, *A Tour Through the Whole Island of Great Britain*, London, 1742, vol. 3, Letter 9.

34. Minute of 12 June 1913 from Churchill (First Lord of the Admiralty) to the prime minister (Herbert Asquith): CHAR 13/22A/126–7. Cf. Churchill's minute of 1913 on a new programme to construct four new airships: CHAR 13/22B/324.

35. The sound mirrors derived from a French design. For a report on their efficiency see Memorandum by the Secretary of State for War to the War Cabinet, 26 April 1918, TNA CAB 24/49.

36. Other mirrors stood further north, at Hartlepool, Seaham and Sunderland. Sunderland's mirror has been restored as a monument.

37. Chain Home Extra Low stations were positioned at Ravenscar and Huntcliff.

38. Major John Cunliffe-Lister died of wounds received in North Africa in 1943. Philip, the younger son, served as a meteorological reconnaissance pilot. After the war's end he did not readjust, and killed himself.

39. For example, at Hereford, Lincoln, Salisbury and Glasgow.

40. Henry Hinchcliffe, *The Stained Glass Windows of Harry Stammers*, Mindelph Press, 2012.

41. Stammers also took St Francis and the birds as a subject in the 1950 window of St Matthias, Stockbridge, South Yorkshire.

42. United States Strategic Air Command had been ready to deploy and engage within six hours since the preceding Tuesday: *Strategic Air Command Operations in the Cuban Crisis of 1962*, Historical Study No. 90, vol. 1, 55.

43. Stephen Twigge and Len Scott, 'The Other Missiles of October: the Thor IRBMs and the Cuban Missile Crisis', *Electronic Journal of International History*, Article 3, 52.

WEST

1. R. F. Light, 'Cricket's forgotten past: a social and cultural history of the game in the West Riding of Yorkshire, 1820–1870', PhD thesis, De Montfort University, 2008.

2. 'Loiner' is a demonym for Leeds's people. No one knows why.

3. West Yorkshire, South Yorkshire, North Yorkshire, Humberside, Lancashire, Greater Manchester, Cumbria.

10: IN BARNSDALE FOREST

1. Barnsdale Bar Little Chef opened in 1981 and closed in September 2012.

2. J. W. Walker (ed.), *Abstracts of the Chartularies of the Priory of Monkbretton*, Yorkshire Archaeological Society Record Series 66, 1924, charter 315, 105–6.

3. Ray Lock (ed.), *The Court Rolls of Walsham le Willows 1303–1350*, Suffolk Records Society XLI, 1998, 269, 271.

4. John Leland, *The Itinerary in or about the Years 1535–1543*, ed. L. T. Smith, London: Bell, 1909, 51.

5. D. Crook, 'Some Further Evidence Concerning the Dating of the Origins of the Legend of Robin Hood', *English Historical Review* 99, 1984, 530–34.

6. A. J. Pollard, *Imagining Robin Hood: The Late-Medieval Stories in Historical Context*, London: Routledge, 2004, 187.

7. J. C. Holt, 'Hood, Robin (*supp. fl.* late 12th–13th cent.)', *Oxford Dictionary of National Biography*, Oxford University Press, 2004; online edn, Jan 2007 [http://www.oxforddnb.com/view/article/13676, accessed 13 July 2017; J. C. Holt, *Robin Hood*, 2nd edn, London:

Thames & Hudson, 1982; R. B. Dobson and J. Taylor, 'Robin Hood of Barnsdale: A Fellow Thou Has Long Sought', *Northern History* 19, 1983, 201–30; R. B. Dobson and J. Taylor, 'The Medieval Origins of the Robin Hood Legend: A Reassessment', *Northern History* 7, 1972, 1–30; R. B. Dobson and J. Taylor, *Rymes of Robin Hood: An Introduction to the English Outlaw*, London: Heinemann, 1976.

8. 'Robin Hood and the Monk'; 'Robin Hood and the Potter'.

9. Pollard, *Imagining Robin Hood*, 2.

10. Ibid., 3.

11. *Gest*: 'adventure', 'story', through Old French from Latin *gesta*, 'deeds'; Stephen Knight and Thomas H. Ohlgren (eds), 'A Gest of Robyn Hode', in *Robin Hood and Other Outlaw Tales*, 2nd edn, Toronto: Medieval Institute Publications, 2000.

12. A section of a song, canto.

13. J. C. Holt, 'Robin Hood, Robin (*supp. fl.* late 12th–13th cent.)', *Oxford Dictionary of National Biography*, Oxford: Oxford University Press, 2004.

14. *Gest*, Fytte 1, 9–10.

15. *Gest*, Fytte 8, 1765.

16. Pollard, *Imagining Robin Hood*, 80.

17. *Gest*, Fytte 1, 81.

18. *Gest*, Fytte 2, 536–7.

19. *Gest*, Fytte 4, 849–50.

20. See Chapter 3.

21. Helen Phillips, 'Forest, Town and Road: The Significance of Places and Names in some Robin Hood Texts', in Thomas Hahn (ed.), *Robin Hood in Popular Culture*, Cambridge: D. S. Brewer, 2000, 197–214.

22. Pollard, *Imagining Robin Hood*, 57.

23. Ibid.

24. Everett Mendelsohn and Helga Nowotny, *Nineteen Eighty-Four: Science Between Utopia and Dystopia*,

Dordrecht: D. Reidel Publishing Company, 1984, 27.

25. Later in the *Gest* they 'walked up under the Sayles, / And to Watlynge-strete': Fytte 4, 833–4.

26. A. H. Smith, *The Place-Names of the West Riding of Yorkshire*, English Place-Name Society, vol. XXXIII, Part Four, Cambridge: Cambridge University Press, 1961, 52.

27. Dobson and Taylor, *Rymes of Robin Hood*, 126.

28. Some editors have seen Watling Street as an error, noting that the correct name should be Ermine Street (or Ryknild Street – see Chapter 3). In fact, this stretch of road has been known locally as both Watling Street and Roman Ridge.

29. A probate record of 1440 gives All Saints. However, double dedications were not unusual, and the fact that in the twelfth century the church was under the patronage of two families, with two rectors, may strengthen the possibility that two dedications ran side by side.

30. *Gest*, Fytte 1, 108.

31. *Gest*, Fytte 7, 1425–8

32. A manor house called 'Plompton Tower' appears on Saxton's Yorkshire map of 1577. A map of the estate made shortly afterwards shows it in the midst of 'Plompton Parke': TNA Manor of Plompton, DL 44/411, m4.

33. John Leland, *J. Lelandi antiquarii de rebus Britannicis Collectanea*, ed. Thomas Hearne, London, 1774, vol. 1, 54. In the later seventeenth century a drawing was made of a cross-decorated stone slab that had purportedly covered Robin Hood's grave at Kirklees. The drawing is ascribed to the physician and antiquary Nathaniel Johnston (the original is in vol. 16 of William Stukeley's MS Diary, Bodleian Library). The slab has disappeared,

but to judge from the drawing it was of a well-known kind, and stylistically a date of *c.* 1300 would fit its making. The slab was intended to form the lid of a stone coffin or cover for a grave. The drawing shows it parked on edge, so by the time it was made the slab had been displaced and any link with its grave had been broken. The image shows at least two names inscribed on the chamfered margins of the stone. These look secondary: inscriptions are sometimes found around the borders of such stones, but they do not take this form and the lettering is several hundred years younger than the slab. All the signs are, then, that the association with Robin Hood was retrospectively contrived in the sixteenth or seventeenth century. This said, since it is unlikely that the inscribed borders would have been legible at distance from which the slab was sketched, it remains possible that the names and anachronistic orthography are the result of annotations by the artist rather than representations of the names as seen.

34. Michael Drayton reported Robin Hood's resting place as 'Kirkeby', which has been interpreted by some as the district of this name that forms part of Pontefract. However, Drayton was describing the grave in Calderdale: after Halifax 'as Calder comes along, / It chanc'd shee in her Course on Kirkeby cast her eye, / Where merry Robbin Hood, that honest thief doth lye' ('Eight and Twentieth Song', 140). The Calder merges with the Aire above Pontefract at Castleford, whereafter the continuation is called the Aire. A phonetic confusion with Kirklees seems likely.

35. For example, H. L. Gee, *Folk Tales of Yorkshire*, London: Nelson, 1952, 48.

36. *Barnsdale Bar Quarry, Norton, South Yorkshire*: Leeds: West Yorkshire Archaeological Services, 2001.

37. Michael Drayton, *Poly-Olbion*, 'Eight and Twentieth Song' (1621), 139.

38. Leland, *Itinerary*, Part 7, 13.

39. Woodland extended into the adjoining township of Gateforth, which to judge from historic field names was already well dissected by clearances in the fifteenth century: Smith, *Place-Names of the West Riding*, 4, 27–8. In 1677 Thomas Baskerville described his journey from Doncaster to Pontefract, passing through Brotherwood north of Wentbridge, and then 'Fraterwood': 'Thomas Baskerville's Journeys in England, Temp. Car. II', in *Historical Manuscripts Commission, Thirteenth Report, Appendix, Part II. The Manuscripts of His Grace the Duke of Portland: Preserved at Welbeck Abbey*, London: HMSO, 1893, 311.

40. Jacques Le Goff, 'The Wilderness in the Medieval West', in *The Medieval Imagination*, trans. Arthur Goldhammer, Chicago: University of Chicago Press, 1985, 52–3.

41. This was also recorded as Barnsdale Lodge on Carey's map of Yorkshire, 1754–1835.

42. Sheffield City Archives SpSt.164/4; Yorkshire Archaeological Society MS 646.

43. Smith, *English Place-Names*, 4, 37, records the eighteenth-century names in Campsall but references distant or antiquarian sources like Wyntoun's *Chronicle* (*c.* 1420) and Leland (1543) to centre Barnsdale in the adjoining township of North Elmsall, where *Barnysdale Ryg* was recorded in 1468. In fact, the township boundary of North Elmsall connects with that of

Campsall on the line of the Street. Barnysdale Ryg was thus just across the road and part of the hilltop cluster.

44. Walker, *Abstracts of the Chartularies of the Priory of Monkbretton*, 107.

45. Pollard, *Imagining Robin Hood*, 64.

46. Ibid.

47. Holt, 'Hood, Robin' http://www.oxforddnb.com/view/article/13676, accessed 13 July 2017.

48. Knight and Ohlgren, *Robin Hood and Other Outlaw Tales*, 80–6.

49. Wemyss MS II, 3453–6

50. F. H. M. Parker, 'Inglewood Forest', *Transactions of the Cumberland and Westmorland Antiquarian and Archaeological Society* 7–10, 1907–10.

51. Court of Common Pleas, 1429, 7 Henry 6.

52. This may be because it came not from a long narrative poem but from a short, proverbial ditty that could be picked up at a single hearing. The occurrence of variant forms with the same rhythm (like 'Robin Hood in Sherwood stood', found in a Lincoln Cathedral manuscript of *c.* 1425 (Lincoln Cathedral Library MS 132)) strengthens the prospect: Thomas H. Ohlgren, *Robin Hood: The Early Poems, 1465–1560. Texts, Contexts and Ideology*, Newark, NJ: University of Delaware Press, 2007, 19.

53. Cf. Adwick le Street: William Page, *The Certificates of the Commissioners Appointed to Survey the Chantries, Guilds, Hospitals, Etc., in the County of York*, Surtees Society 91, 1894, I, 174.

54. A. J. Pollard, 'Idealising Criminality: Robin Hood in the Fifteenth Century', in *Pragmatic Utopias: Ideals and Communities, 1200–1630*, eds. Rosemary Horrox and Sarah Rees Jones, Cambridge: Cambridge University Press, 2011, 156–73, at 162.

55. Dobson and Taylor, *Rymes of Robin Hood*, 21.

56. Spellings, rhymes and other textual indications put composition of the *Gest* in the north, and very possibly in Yorkshire: M. Ikegami, 'The Language and the Date of *A Gest of Robyn Hode*', *Neuphilologische Mitteilungen* 3/96, 1995, 271–81.

57. For example: 'some of them muster in Barnsdale and Barnslay', Sir Brian Hastyngs to Shrewsbury, 17 October 1536 (*Letters and Papers, Foreign and Domestic, of the Reign of Henry VIII*, vol. 11, ed. J. Gairdner, London: HMSO, 1888, no. 759); '"To this he saith" on the Tuesday he went with the "vaward", about 12,000, to a plain above Barnesdale nigh Hampall': examination of Robert Aske, 11 May 1537 (*Letters and Papers*, vol. 12, no. 1175); 'That day they went forward towards Doncaster, and, beside Barnesdale, came Lancaster herald with a letter from the lords at Doncaster': confession of William Stapulton, 1537 (*Letters and Papers*, vol. 12, no. 392).

58. Michael Wood, 'Searching for Brunanburh: The Yorkshire Context of the "Great War" of 937', *Yorkshire Archaeological Journal* 85, 2013, 158.

59. See note 29.

60. David Crouch, 'The Origin of Chantries: Some Further Anglo-Norman Evidence', *Journal of Medieval History* 27, 2001, 159–80, at 161.

61. *Valor Ecclesiasticus*, Record Commission 1810–1834, v.179.

62. Richard Campsall was born at Campsall probably between 1280 and 1285. By 1322 he was master in theology at Oxford and was subsequently a senior office bearer in Merton College: E. A. Synan, *The Works of Richard Campsall*, 2

vols., Toronto: Institute of Pontifical Studies, 1982.

63. Page, *Certificates of Commissioners Appointed to Survey the Chantries*, 1, 200.

64. A writ of proof of age of Edward Hastynges, knight, stated that he had been baptised in this chapel in 1382: 27 May 1403, *Calendar of Inquisitions Post Mortem* 18 Part IX, C 137/40 no. 43.

65. Page, *Certificates of Commissioners Appointed to Survey the Chantries*, 1, 164–5; *As You Like It* V.4, 'You and you are sure together / As the winter to foul weather'.

66. Stacey Gee, '"At the Sygne of the Cardynalles Hat": the Book Trade and the Market for Books in Yorkshire, *c.* 1450–1550', unpublished PhD thesis, University of York, 1999.

67. Page, *Certificates of Commissioners Appointed to Survey the Chantries*, 1, 175–82.

68. Leland, *Itinerary,* Part 4, 14–15.

69. Page, *Certificates of Commissioners Appointed to Survey the Chantries*, 1, 200; Leland, *Itinerary,* Part 4, 14.

70. Introduction of 1483 to statutes of Jesus College: Arthur Francis Leach, *Early Yorkshire Schools*, Leeds: Yorkshire Archaeological Society, 1903, Record Series 33, vol. II, 109–30.

71. Page, *Certificates of Commissioners Appointed to Survey the Chantries*, 1, 201.

72. James Willoughby, 'The Educational Patronage of Thomas Rotherham, Archbishop of York (1423–1500): The Evidence of Incunabula Once at Jesus College, Rotherham', *Transactions of the Cambridge Bibliographical Society* 14:4, 2011, 293–316. The book list is printed in Leach, *Early Yorkshire Schools*, vol. II, 160–67.

73. Tuning pegs of harps or fiddles found at the College of Vicars Choral in York conjure a world of off-duty minstrelsy: Nicola Rogers, '"Wine, Women and Song": Artefacts from the Excavations at the College of Vicars Choral at the Bedern, York', in *Cantate Domino: Vicars Choral at English Cathedrals. History, Architecture and Archaeology*, eds. Richard Hall and David Stocker, Oxford: Oxbow, 2005, 164–87, esp. 168–9.

74. The school was refounded three years later as Rotherham Grammar School.

75. Some portions of its pioneering brickwork survive encased in later buildings: J. A. Wight, *Brick Building in England from the Middle Ages to 1550*, London: J. Baker, 1972, 136–53; Historic England List entry number 1192176.

76. Nikolaus Pevsner, *Yorkshire: The West Riding* (*The Buildings of England*), revised by Enid Radcliffe, Harmondsworth: Penguin Books, 1967, 518.

77. Historic England List entry number 1287066.

78. Rotherham's brick-building elsewhere included the Old Schools at Cambridge, a range at the archiepiscopal palace at Bishopthorpe, possible contributions to the palaces at Scrooby and Cawood, and to the Archdiocese of York's London houses at Battersea and York Place: David H. Kennett, 'Thomas Rotherham, a Fifteenth-Century Bishop and Builder in Brick: A Preliminary Note', *British Brick Society Information* 112, 2010, 6–18.

79. *Testamenta Eboracensia*, vol. 4, Surtees Society vol. 53, 1869, 138–48, at 140.

80. E. Bateson, 'Notes on a Journey from Oxford to Embleton and Back in

1464', *Archaeologia Aeliana*, new series
16, 1894, 113–20, at 118–19. Merton
College accounts show similar
journeys in 1331: G. H. Martin,
'Road Travel in the Middle Ages:
Some Journeys by the Warden and
Fellows of Merton College, Oxford,
1315–1470', *Journal of Transport
History*, new series 3, 1976, 159–78, at
175–6.

11: MUNGO, SHODDY, AND ENEMIES WITHIN

1. J. B. Priestley, *Margin Released: A
   Writer's Reminiscences and Reflections*,
   New York: Harper and Row, 1962,
   140.
2. 'Original Correspondence', *Bradford
   Observer*, 17 July 1834.
3. 'Death of Richard Oastler', *Standard*,
   23 August 1831.
4. Parliamentary Papers 1831–1832, xv,
   454–5.
5. *Bradford Observer*, 31 July 1834, 1.
6. The Factory Act 1847 stipulated a
   maximum of sixty-three hours per
   week for women and young people
   aged 13–18, reduced to fifty-eight
   hours from the following year.
7. *Bradford Observer*, 29 August 1861, 5.
8. *Huddersfield Chronicle*, 31 August
   1861.
9. Roger Jeness, 'Landscape and the
   Geographical Imagination of J. B.
   Priestley: 1913–1930, PhD thesis,
   University of Sussex, 2013, 40.
10. At this point, Hill was co-owner. He
    became sole proprietor in 1916.
11. T. Sheppard, *Yorkshire's Contribution
    to Science, with a Bibliography
    of Natural History Publications*,
    London: A. Brown, 1916, 8; J.
    James, *The History and Topography
    of Bradford*, London/Bradford:
    Longman, Brown, Green, and
    Longmans/Charles Stanfield, 1841,

19; W. Scruton, *Pen and Pencil
Pictures of Old Bradford*, Bradford:
Thomas Brear, 1889, 107–8; J.
B. Morrell, 'Wissenschaft in
Worstedopolis: Public Science in
Bradford, 1800–1850', *British Journal
of the History of Science* 18, 1985, 1–23,
at 2.

12. John Ruskin, from lecture 'Traffic'
    given in Bradford, 21 April 1864,
    published in *The Crown of Wild
    Olive*, London: George Allen
    & Sons, 1906, 97; cf. Morrell,
    'Wissenschaft in Worstedopolis', 2.
13. House of Commons debates 22
    and 28 June 1916, Hansard, vol. 83
    cc.349–50 and cc.843–4.
14. *Yorkshire Post*, 18 January 1936.
15. Ibid., 24 January 1918.
16. Susan Duxbury-Neumann, *Little
    Germany: A History of Bradford's
    Germans*, Stroud: Amberley, 2015,
    16–17.
17. J. B. Priestley, *Bright Day*, London:
    Heinemann, 1946, 81.
18. John Baxendale, *Priestley's England:
    J. B. Priestley and English Culture*,
    Manchester: Manchester University
    Press, 2007, 39; Priestley, *Bright Day*,
    43–4; Priestley, *Margin Released*,
    28–9.
19. Duxbury-Neumann, *Little Germany*,
    60–109.
20. For discussion of these see Jeness,
    'Landscape and the Geographical
    Imagination of J. B. Priestley', 38–57.
21. 'Shock city': Asa Briggs, *Victorian
    Cities*, London: Odhams Press,
    1963, 51. 'Certainly there has been
    a march of the mind from the
    city. One remembers Fred Delius,
    Rothenstein, Sir Edward Appleton,
    J. B. Priestley, John Braine,
    Barbara Castle, Alan Bullock, Vic
    Feather': Morrell, 'Wissenschaft in
    Worstedopolis', 2.
22. J. B. Priestley, *The Good Companions*,
    London: Heinemann, 1929, 2.

23. Ted Hughes, 'Moors', in *Remains of Elmet: New Selected Poems 1957–1994*, London: Faber & Faber, 1995, 160.

24. Peter of Pomfret is a character in Shakespeare's *King John* (IV.2).

25. For example, by the Nidd at Knaresborough (St Robert), beside a tributary of the Don (Richard Rolle, the hermitage at Conisbrough), the tunnelling hermits at Pontefract on the Aire, or in the twelfth century further up the Aire valley at Kirkstall.

26. See Chapter 7.

27. Ted Hughes, *Paris Review* 134, Spring 1995.

28. Sources for chapel-going 'Cleckleywyke', another Priestley composite, in *When We are Married* (1938).

29. J. B. Priestley, *English Journey*, London: Heinemann, 1934, 4.

30. Briggs, *Victorian Cities*, 33–34.

31. O. L. Gilbert: *The Flowering of the Cities: The Natural Flora of 'Urban Commons'*, Peterborough: English Nature, 1992; *The Ecology of Urban Habitats*, London: Chapman and Hall, 1989.

32. Steve Ely, *Ted Hughes's South Yorkshire: Made in Mexborough*, London: Palgrave Macmillan, 2015.

33. See p. xx.

34. Emma Priestley died in 1896. Jonathan remarried two years later, to Amy Fletcher (1868–1935).

35. Simon Taylor and Kathryn Gibson, *Manningham: Character and Diversity in a Bradford Suburb*, Swindon: English Heritage, 2010, 1.

36. See Chapter 9.

37. Philip was born Lloyd-Greame. He took the family name when his wife inherited the estate after the death of Samuel Lister's son.

38. *Manchester Guardian*, 2 January 1891, 11.

39. *Leeds Mercury*, 4 February 1891.

40. *North-East Daily Gazette*, 14 April 1891.

41. Derek Barker, 'The Manningham Mills Strike, 1890–91: 'Low wages, good water, and no unions', *Northern History* 50:1, 2013, 93–114; Cyril Pearce, *The Manningham Mills Strike, Bradford: December 1890–April 1891*, Hull: University of Hull, 1975; Keith Laybourn, 'The Manningham Mills Strike, December 1890 to April 1891', in David James, Tony Jowitt and Keith Laybourn, eds., *The Centennial History of the Independent Labour Party*, Halifax: Ryburn Publishing, 1992, 117–36.

42. Margaret Thatcher Foundation, notes for speech to 1922 Committee, 19 July 1984.

43. The National Coal Mining Museum for England, Caphouse Colliery, Overton, Wakefield.

12: ALL FOR YOUR DELIGHT

1. The camp station opened in 1947 and closed in July 1977.

2. T. E. Pritt, *Yorkshire Trout Flies*, Leeds: Goodall and Suddick, 1885.

3. T.E. Pritt, *The Book of the Grayling*, Leeds: Goodall and Suddick, 1888.

4. Tom Bradley, 'Yorkshire Rivers: No. 9, The Aire', *Yorkshire Post*, 1893, 35–6.

5. F. M. Walbran, *Grayling and How to Catch Them*, Scarborough: The Angler Co., 1895.

6. 'Mr Francis M. Walbran Drowned While Fishing', *Fishing Gazette*, 20 February 1909, 142.

7. '"Max" Walbran: Some Stories and an Appreciation in the *Yorkshire Evening Post*', *Fishing Gazette*, 27 February 1909, 168.

8. Oral history recording cited by D. H. Whiteley, 'Ancestry, Boardmaking and Commentary', n.d., 106–7.

9. *Fishing Gazette*, 27 January 1917.

10. Eilert Ekwall, *English River-names*, Oxford: Clarendon Press, 1968, 99.

11. Geddes was appointed in May 1919.

12. Kathryn A. Morrison and John Minnis, *Carscapes: The Motor Car, Architecture and Landscape in England*, New Haven, Conn., and London: Yale University Press, 2012, 245.

13. Petrol was cheapest in 1928; the price dipped again in 1931-2: Joe Hicks and Grahame Allen, *A Century of Change: Trends in UK Statistics since 1900*, House of Commons Research Paper 99/111.

14. Andrew Jenkinson, *Caravans: The Illustrated History, 1919-1959*, Dorchester: Veloce, 2003.

15. Sandra Dawson, 'Working Class Consumers and the Campaign for Holidays with Pay', *Twentieth Century British History* 18:3, 2007, 277-305.

16. John K. Walton, 'The Demand for Working-Class Seaside Holidays in Victorian England', *Economic History Review*, new series, 34:2, 1981, 249-65, at 256.

17. Ibid.

18. Ibid., 259-60.

19. Cited in Margaret Drabble, 'Filey', in *Edge of Heaven: The Yorkshire coast*, ed. Lee Hanson, Ilkley: Great Northern Books, 2011, 129-44, at 134-5.

20. See Introduction.

21. World War 1 Appeal Tribunal Papers, North Yorkshire Archive, NRCC/Cl9/2/1.

22. J. W. Graham, *Conscription and Conscience: A History 1916-1919*, London: George Allen and Unwin, 1922, 240-41.

23. Ibid., 237.

24. 'The Letters of an Englishman', *Daily Mail*, 28 April 1917, 2.

25. 'Cushy Jobs for Unworthy Creatures', *Daily Mail*, 16 April 1917, 4.

26. About 10,500 civilian males were interned in September 1914. Restrictions were eased in the following months, but reimposed with greater force after the sinking of the *Lusitania* in May 1915. Lofthouse Park was one of the main centres of civilian internment: Panikos Panayi, *The Enemy in Our Midst: Germans in Britain during the First World War*, Oxford: Oxford University Press, 1991; Panayi, *Prisoners of Britain: German Civilian and Combatant Internees during the First World War*, Manchester: Manchester University Press, 2012; Paul Cohen-Portheim, *Time Stood Still: My Internment in England, 1914-18*, London: Duckworth, 1931.

27. 'The Revolting "Conchie"', *Daily Mail*, 9 October 1917, 2.

INDEX